The
Discipline
of
Market
Leaders

THE
DISCIPLINE
OF
MARKET
LEADERS

CHOOSE YOUR CUSTOMERS,
NARROW YOUR FOCUS,
DOMINATE YOUR MARKET

MICHAEL TREACY AND
FRED WIERSEMA

Addison-Wesley Publishing Company

Reading, Massachusetts   Menlo Park, California   New York   Don Mills, Ontario
Wokingham, England   Amsterdam   Bonn   Sydney   Singapore   Tokyo
Madrid   San Juan   Paris   Seoul   Milan   Mexico City   Taipei

Many of the designations used by manufacturers and sellers to distinguish their products are claimed as trademarks. Where those designations appear in this book and Addison-Wesley was aware of a trademark claim, the designations have been printed in initial capital letters.

Library of Congress Cataloging-in-Publication Data

Treacy, Michael
    The discipline of market leaders : choose your customers, narrow
your focus, dominate your market / by Michael Treacy, Fred Wiersema.
       p. cm.
    Includes index.
    ISBN 0-201-40648-9
    1. Competition. 2. Market segmentation. 3. Quality of products.
4. Customer service. I. Wiersema, Fred. D. II. Title.
HD41.T67   1995
658.8—dc20   94-34808
                                          CIP

Jacket design by Jean Seal
Text design by Ken Silvia Design Group
Set in 12-point Adobe Garamond by Cambridge Prepress Services

4 5 6 7 8 9-MA-9998979695
Fourth printing, March 1995

Addison-Wesley books are available at special discounts for bulk purchases by corporations, institutions, and other organizations. For more information, please contact the Corporate, Government, and Special Sales Department, Addison-Wesley Publishing Company, Reading, MA 01867, 1-800-238-9682.

There is only one leader.

— Philip H. Knight,
chairman and CEO, Nike

To my wife, Evelyn,
who is the center of my life.
— M.T.

To Catherine and Annelise,
whose love I cherish more than anything.
— F.W.

CONTENTS

ACKNOWLEDGMENTS

The central themes of this book have their roots in our consulting work with dozens and dozens of corporations, both large and small. Some of these corporations were market laggards that succeeded in changing their course, others are market innovators that are remarkable for their sustained market leadership. While these companies, and the many outstanding executives that manage them, must go unnamed, we are deeply indebted to them for inspiring us to think in new and ever-evolving ways.

Our insights were greatly enriched by our dialogue with executives from the 65 companies that over the past seven years sponsored the CSC Index Alliance research program. We thank them for their practical input. We thank the program's director, Jay Michaud, for being our steady thinking partner. We are also indebted to Ron Christman and Joe Ferreira for their unswerving support of our work.

Our gratitude goes to the many professionals at CSC Index who helped us articulate the discipline of market leaders. We would like to single out a few who went the extra mile (without taking away from the gratitude we feel for the many others whose names aren't included here): Michael Black, Russell Brackett, Bob Buday, Phil Casciotti, Allan Cohen, Martha Craumer, Bob Dantowitz, David Robinson, Rich Fagan, Elizabeth Gorman, Jim Hall, Steve Hoffman, Jim Kennedy, Phil Lawrence, Tom Manning, Anne Miller, Rhoda Pitcher, Roger Pratesi, Brad Power, Judi Rosen and Greg Tucker.

To Tom Waite goes a very special thanks for guiding the overall development of this book and for his contributions to its content. He was instrumental in moving this book from vision to reality.

This work and our views on business and competition have been influenced by several academicians including Bob Buzzell, George Day, John Henderson, Jack Rockart, Lou Stern, and N. Venkatraman. To each we extend our thanks.

We particularly want to thank Bill Birchard, Donna Sammons Carpenter, Robie Macauley, Tom Richman, and the other talented writers, editors, and researchers at Wordworks, Inc. — Susan Buchsbaum, Maurice Coyle, Martha Lawler, Mike Mattil, Cindy Sammons, Charles

Simmons, and Sebastian Stuart. Our thanks also to Helen Rees, our literary agent who offered guidance at every step; and William Patrick, our editor at Addison-Wesley whose support and patience were always there.

Finally, our appreciation goes to Lore Caturano, Leeann Falzone, Lyn Goldman, Amy Reuss, and Sue Walseman, who carried the burden of keeping us organized and productive. Without them, nothing would get done.

INTRODUCTION

Is your company willing to cannibalize its hottest product with a risky, untested new one? Offer a service at a loss hoping to establish a long-term relationship? Link up with an adversary to drive its costs even lower?

If not—or if you believe the answer isn't of paramount importance—get used to mediocre market performance and to playing competitive catchup continuously. Your company will never be a market leader. Not until it learns discipline.

This is a book about the discipline needed to become and remain number one. It's based on five years of research and practice conducted with CSC Index, the fastest-growing management consulting firm in the world. A business revolution was launched in 1993 with the publication of *Reengineering the Corporation*, by Michael Hammer and James Champy. That book, a bestseller around the globe, advanced a set of ideas about how companies can redesign the way they do their work. This book, *The Discipline of Market Leaders*, will change the way business leaders think about what work their companies should do. *Reengineering the Corporation* was about how to run a race. *The Discipline of Market Leaders* is about choosing the race to run.

In today's economic environment, you've got to reinvent the rules of competition. Some familiar companies with long records of success—Wal-Mart, Hewlett-Packard, Sony—have already learned to play by the new rules. You'll read about those companies, as well as some newcomers—such as Cott Corporation—that are on their way to becoming market leaders.

These companies, both old and new, are redefining business competition in one market after another. By relentlessly driving themselves to deliver extraordinary levels of distinctive value to carefully selected customer groups, these market leaders have made it impossible for other companies to compete on the old terms. *The Discipline of Market Leaders* also looks at how some of the world's best-known companies—Digital Equipment Corporation, Eastman Kodak, and International

Business Machines—have failed to recognize and adapt to this new competitive reality. Their stumbling performances provide cautionary tales about survival, to which close attention must be paid.

To whom is this book useful? It offers critical insights to the team struggling to turn around IBM. It offers market leaders, such as Home Depot, Nike, and McDonald's, paths forward for sustaining their leadership. It applies to someone managing a small business—as small as a corner donut shop in Anytown, U.S.A.—by highlighting the key issues upon which that business will succeed or fail.

The message of *The Discipline of Market Leaders* is that no company can succeed today by trying to be all things to all people. It must instead find the unique value that it alone can deliver to a chosen market. Why and how this is done are the two key questions the book addresses.

When we talk about companies in these pages, we are actually addressing the management of business units. We ask that the reader understand that this simplifies our task: to introduce, define, and develop three concepts that every business unit will find essential. The first is the value proposition, which is the implicit promise a company makes to customers to deliver a particular combination of values—price, quality, performance, selection, convenience, and so on. The second concept, the value-driven operating model, is that combination of operating processes, management systems, business structure, and culture that gives a company the capacity to deliver on its value proposition. It's the systems, machinery, and environment for delivering value. If the value proposition is the end, the value-driven operating model is the means. And the third concept, which we call value disciplines, refers to the three desirable ways in which companies can combine operating models and value propositions to be the best in their markets.

We have identified three distinct value disciplines, so called because each discipline produces a different kind of customer value. Choosing one discipline to master does not mean that a company abandons the other two, only that it picks a dimension of value on which to stake its market reputation.

The first value discipline we call *operational excellence.* Companies that pursue this are not primarily product or service innovators, nor do they cultivate deep, one-to-one relationships with their customers. Instead, operationally excellent companies provide middle-of-the-market products at the best price with the least inconvenience. Their proposition to customers is simple: low price and hassle-free service. Wal-Mart epitomizes this kind of company, with its no-frills approach to mass-market retailing.

The second value discipline we call *product leadership*. Its practitioners concentrate on offering products that push performance boundaries. Their proposition to customers is an offer of the best product, period. Moreover, product leaders don't build their positions with just one innovation; they continue to innovate year after year, product cycle after product cycle. Intel, for instance, is a product leader in computer chips. Nike is a leader in athletic footwear. For these and other product leaders, competition is not about price; it's about product performance.

The third value discipline we have named *customer intimacy*. Its adherents focus on delivering not what the market wants but what specific customers want. Customer-intimate companies do not pursue one-time transactions; they cultivate relationships. They specialize in satisfying unique needs, which often only they, by virtue of their close relationship with—and intimate knowledge of—the customer, recognize. Their proposition to the customer: We have the best solution for you—and we provide all the support you need to achieve optimum results and/or value from whatever products you buy. Airborne Express, for example, practices customer intimacy with a vengeance, achieving success in a highly competitive market by consistently going the extra mile for its selectively chosen customers.

A company's choice of the value discipline it will pursue to market leadership is not arbitrary. It's the result of extensive inquiry and analysis—of the company and of its market. *The Discipline of Market Leaders* will show you how to undertake this process of choosing.

One point deserves emphasis: Choosing to pursue a value discipline is not the same as choosing a strategic goal. A value discipline can't be grafted onto or integrated into a company's normal operating philosophy. It is not a marketing plan, a public relations campaign, or a way to chat up stockholders. The selection of a value discipline is a central act that shapes every subsequent plan and decision a company makes, coloring the entire organization, from its competencies to its culture. The choice of value discipline, in effect, defines what a company does and therefore what it is.

In *The Discipline of Market Leaders*, we show how companies deliver continually-improving value to the customers they've elected to serve. We explain why market leadership demands adherence to a value discipline and how companies can structure themselves to succeed in that pursuit. Some of the world's most successful companies have become number one in their field in this way. Our book is the first to identify, codify, and illuminate key aspects of their winning strategies.

We draw upon our experience in working with real companies facing real problems—and upon the results of a comprehensive, three-year study of more than 80 corporations in more than three dozen markets. We collected data from these market leaders as well as from companies whose recent results have been less than stellar. The objective was to account for one group's success and the other's failure to achieve and sustain market leadership.

In the chapters ahead, we dissect the critical practices of companies that have pursued an operating model to reach the top. We then extract examples of the sharpest thinking in business today. What emerges is a clear picture of the kinds of companies that will be tomorrow's stars, and the strategies that will place them in the firmament.

Many of the ideas you encounter in *The Discipline of Market Leaders* will surprise you. That's because they seem out of step with the current, widely-held notion that to identify core competencies and to reengineer a company's business processes is to assure its competitive future. The flaw in this thinking lies not with the principles and practices of core competencies and reengineering, which are powerful concepts and tools. The flaw lies in the assumption that they are all that an ailing company needs. They are not. Sick or not, if a company is going to achieve and sustain dominance, it must first decide where it will stake its claim in the marketplace and what kind of value it will offer to its customers. Then it can identify core competencies and reengineer the processes that make up the operating model required to get the job done.

In this book, we will describe why, in today's new competitive environment, the choice of a value discipline must be made; we will provide help in identifying the right choice; and we will detail the means for implementing it. In other words, we will do everything but make the choice for you.

That's your job.

1

How to Fail in Business Without Even Trying

CHAPTER 1

HOW TO FAIL IN BUSINESS WITHOUT EVEN TRYING

■ Why is it that Casio can sell a calculator more cheaply than Kellogg's can sell a box of cornflakes? Does corn cost that much more than silicon?

■ Why is it that it takes only a few minutes and no paperwork to pick up or drop off a rental car at Hertz's #1 Club Gold, but twice that time and an annoying name/address form to check into a Hilton hotel? Are they afraid you'll steal the room?

■ Why is it that you can get a bright, cheerful, informed L.L. Bean salesperson on the phone to take your catalogue order, but after spending much more money on premiums, you can hardly get an Aetna health-insurance claims rep to answer your call? Don't they know you have choices?

■ Why is it that FedEx can "absolutely, positively" deliver a package overnight, but Delta, American, and United Airlines have trouble keeping your bags on your plane? Do they think you don't care?

■ Why is it that Swatch can produce an inexpensive watch that will run accurately for years, but Rolex can't or won't do the same for a hundred times the money? Doesn't Rolex understand that the standards of Swiss reliability have changed?

■ Why is it that a great cup of coffee is available for $1 at Starbucks but not from the airport concessions that sell the decades-old Maxwell House brand from General Foods? Is good coffee a secret formula?

■ Why is it that you can buy intuitive, engaging, and intelligent software for a Sega video game, but after a decade on the market, Lotus 1-2-3 has basically the same look, feel, and function? Do the people at Lotus believe they can rest on their laurels forever?

■ Why is it that Lands' End remembers your last order and your family members' sizes, but after 10 years of membership, you are still being solicited by American Express to join? Don't the people at AmEx know that you're a customer?

■ Why is it you can get patient help from a Home Depot clerk when selecting a $2.70 package of screws, but you can't get any advice when purchasing a $2,700 personal computer from IBM's direct ordering service? Doesn't IBM think customer service is worth its time?

Why is it that some companies endear themselves to us while others just don't seem to know how to please? Don't the latter see what they are doing—and not doing? How long do they think they can get away with it?

No one goes to work in the morning intending to fail. But managers at a great many companies, for all practical purposes, have chosen failure. Don't they see that the world has changed?

Customers today want more of those things they value. If they value low cost, they want it lower. If they value convenience or speed when they buy, they want it easier and faster. If they look for state-of-the-art design, they want to see the art pushed forward. If they need expert advice, they want companies to give them more depth, more time, and more of a feeling that they're the only customer.

Whatever customers want today, they want more of it. That's precisely why companies like Kellogg, Kmart, Hilton, Aetna, American, United, Delta, Rolex, General Foods, Lotus, General Motors, and IBM are on a slippery slope. One or more companies in their markets have increased the value offered to customers by improving products, cutting prices, or enhancing service. By raising the level of value that customers expect from everyone, leading companies are driving the market, and driving competitors downhill. Companies that can't hold their own will slip off a cliff.

Today's market leaders understand the battle in which they're engaged. They know they have to redefine value by raising customer expectations in the one component of value they choose to highlight. Casio and Wal-Mart, for instance, establish new affordability levels for familiar products; Hertz makes car rental nearly as convenient as taking a cab; L.L. Bean proves that telephone sales reps can offer service as friendly as a good neighbor's; Swatch reinvents the concept of Swiss reliability;

Sega makes real what was barely imagined; Lands' End shows individuals that they're not just a number; and Home Depot proves that old-fashioned, knowledgeable advice hasn't gone the way of trading stamps.

But wait a minute. These companies don't shine in every way. Wal-Mart doesn't peddle haute couture; L.L. Bean doesn't sell clothing for the lowest possible cost; and Starbucks doesn't slide a cup of java under your nose any faster or more conveniently than anyone else. Yet all of these companies are thriving because they shine in a way their customers care most about. They have honed at least one component of value to a level of excellence that puts all competitors to shame.

THE PRICE IS RIGHT

Price is one of those components. Not long ago, companies raised prices as if by reflex. The standard rule of the road: When your costs escalate, raise your prices to protect your margins and earnings. The rule today: Lower your prices by this afternoon and lower them again tomorrow. The tight lid on pricing demands that firms get a grip on costs.

Woe to Mercedes Benz. It built its image on high-quality, high-performance products, but has found itself losing ground to the likes of Lexus and Infiniti. Mercedes has shifted some assembly to South Korea and Mexico and is planning to spend over half-a-billion dollars for a Spanish minivan plant. These moves will keep it in the luxury car game, but they are only a start.

Woe to Boeing, which staked its reputation on product design. With Airbus gnawing at its market share and emaciated airlines too poor to afford its flagship planes, Boeing is streamlining operations. Boeing CEO Frank Shrontz, with reference to GM and IBM, warned: "It could happen to us, if we don't do things differently." Mandating across-the-board efficiencies, Boeing expects to cut costs 25 percent over the next few years.

Woe to Coca-Cola and Nabisco. Private-label competitors, pricing products 30 to 40 percent lower, are starting to eat the big guys' lunch. The President's Choice label of Loblaw, the Canadian grocery chain, now accounts for 55 percent of cola sales in its own stores and sells throughout the U.S. in chains like D'Agostino and Star Market. Loblaw chocolate-chip cookies are giving Nabisco's Chips Ahoy, a long-time market leader, a drubbing. Loblaw underprices Nabisco while baking cookies with pure butter and 38 percent chocolate. Nabisco scrapes by with vegetable shortening and only 24 percent chocolate. Ahoy, Nabisco. Your customers won't need MBAs to taste the difference.

So much for the price protection that reputation, brand, or patent has afforded. In many markets, the only established way to improve value to customers is to cut prices. If you haven't started thinking about cutting your way to leanness, it's going to cost you later.

BUYING TIME

Time is another component of value. A decade's worth of technological advances have repriced the value of time and, in turn, reset customers' expectations. Customers now penalize suppliers that infringe on their time, whether through delays, mistakes, or inconveniences.

Delays set customers' teeth grinding. People are tired of waiting around for service. They fell head over heels for the fast-food, drive-through lifestyle long ago. Now they find even that's too sleepy. Today's customers demand operations that are airborne, on-line, and real-time. "Soon" is not the answer they want to hear when they ask, "When?" They count speed of response as a key value dimension. Their directive to the marketplace: Continuously shrink the interval between our need and when you fill it.

Some retailers may wake up too late to heed the sound of customers' fists pounding on the counters for faster service. If store managers don't reform, customers will shorten checkout lines themselves by shopping elsewhere. Supermarket chains such as Kroger, A&P, ShopRite, and Publix are already experimenting with self-service checkout systems, where customers scan items themselves and pay by credit card. Sure, some customers may miss the face behind the counter. And others may hesitate to perform the clerical work themselves. But rest assured, many will make these small sacrifices to save time.

Everybody makes mistakes. That is, they used to. But the more customers value their time, the less they tolerate screw-ups. The more error-free products and services become, the higher customers' expectations. Customers are taking their cue from the likes of FedEx. FedEx's service guarantee is a shining example of what is possible and, therefore, what customers feel they can demand. They view flawless service as the norm, not the exception. They have been spoiled by catalogue marketers like L.L. Bean and Lands' End, which zealously pursue zero-defect service. They believe there's no excuse for lousing up a customer order. Pity suppliers that don't keep up with these expectations—their customers will walk, or make them pay for being sloppy. Wal-Mart, for example, started to charge suppliers for the cost of overly common mistakes, such as inaccurate invoices, incomplete deliveries, and late shipments.

Inconveniences similarly wear on customers' patience. If one company doesn't make it easier for the customer to do business, another will. Round-the-clock availability, as offered by automatic teller machines, that give you access to your money any time, anywhere, in this country or abroad, are setting the pace. Charles Schwab and AT&T conduct business at the times most convenient to their customers, not just to themselves.

Today, companies bring service to the customer, not the other way around. Mail order, growing swiftly, successfully sells both mundane and high-ticket, high-tech products such as personal computers over the phone. Auto glass businesses replace windshields wherever customers have parked their cars. Even gourmet restaurants—not just pizza houses—deliver meals in some cities to whatever home or office the customer specifies. Are managers in other industries watching? Are lubrication and auto-repair businesses ready to follow the windshield businesses' lead? If the pizza industry can deliver two sodas and a pizza to your doorstep for $8, why can't somebody crack the code on home delivery of groceries?

Customers are demanding back the time that transactions once took away. If companies don't let customers buy time in every transaction, if that's not part of what they're selling, customers will have little time for them—they'll go to someone else who does it faster.

EXPECTING THE UNEXPECTED

Premium service is another component of value. Not long ago, only special customers felt entitled to special treatment. But today the extraordinary is becoming ordinary. What was once premium service—speedy, flawless, and responsive—is commonplace. Customers now expect their suppliers to go beyond the expected. Truly premium services are being redefined.

IBM and DEC are finding that many customers don't want to buy just equipment. They want to buy the services of a supplier that designs, builds, installs, operates, upgrades and troubleshoots whole networks of computers. Sometimes they want to outsource the whole shebang. If that's not what they get, they'll pick up the phone to call the likes of EDS and CSC, which offer all-inclusive systems integration and outsourcing services. In shopping for transportation, customers don't want just cheaper trucking rates, they want someone to take their shipping problems away along with their freight. Companies like Roadway Logistics Systems and Consolidated Freight have responded. Yellow Freight, playing catch-up, risks losing some of its best customers. In retailing, many consumers obviously want more than just the lower prices and hassle-free operations of power retailers and so-called category killers. Nordstrom and Home Depot are proving that customers value expert shopping advice, too.

WHEN GOOD ISN'T GOOD ENOUGH

Yet another component of value is quality. In years past, quality was something you could add to a product as an extra. Now it's a given in all products. High quality is the cost of admission to the market. Without it, you're not even in the ballpark.

For many people, quality means new features; it used to be that you could add a bell or whistle to an old product and call it improved. Now, if you don't revamp the product to reflect the latest technology, nobody wants it. In the first half of this century, German cameras and lenses were unequaled in workmanship and popularity, prized by all serious photographers, professional and amateur. Suddenly, in the 1950s, a Japanese company named Nikon introduced lenses made from rare-earth glasses. Nikon's new products astonished photographers everywhere and spurred the company's sales. Japanese companies went on to pioneer the single-lens-reflex camera, now standard equipment in 35-millimeter photography, as well as other innovations such as automatic focusing. As one feature after another tumbled off the drawing board, Japanese camera makers whipped all overseas competitors. Today, competition remains heated among the Japanese, but between the Japanese and others it hardly exists.

The ability to generate high-quality, low-cost products that the market has never seen before is the lynchpin of Japan's corporate growth.

Over the past eight years, Canon, Honda, and NEC have all grown 200 to 300 percent, a phenomenal amount. How do they do it? They create and market new and superior products better than anyone else. Until recently, Kodak was the uncontested worldwide leader in film; today Fuji is in hot pursuit. No one thought it could happen.

Through it all, some companies never got the point about quality. They not only declined to improve their products, they degraded them, and have been penalized as a consequence. Kraft General Foods took its consumers for granted and cheapened the beans in Maxwell House coffee. Many once-loyal customers looked elsewhere for their coffee fix and abandoned Maxwell House down to the last drop.

THE NEW WORLD OF COMPETITION

Four new premises underlie successful business practice today:

■ Companies can no longer raise prices in lockstep with higher costs; they have to try to lower costs to accommodate rising customer expectations.

■ Companies can no longer aim for less than hassle-free service. Their customers enjoy effortless, flawless, and instantaneous performance from one industry and want it from every other.

■ Companies can no longer assume that good basic service is enough; customers demand premium service—and raise their standards continuously.

■ Companies can no longer compromise on quality and product capabilities. They must build products to deliver nothing less than superiority and eye-popping innovation.

In the buyer-seller relationship, the tables have turned—*caveat* vendor, the buyer is king. Look no farther than the computer industry. In the beginning of the PC age, the market had room for legions of new, enterprising companies. But the party ended once customers figured out that offering state-of-the-art technology was simply no longer, well, state-of-the-art. Most computer companies peddled pretty much the same technology, supplied by product leaders such as Intel, Microsoft, and Seagate. Customers wanted more, and then still more, and had the clout to insist on better deals.

Today, computer makers offer such services as toll-free hot lines, on-site service, automated fax response systems, and overnight delivery of replacement parts. Why? Because whether they sell $2,500 home computers or $25 software packages, customers demand that PC equipment suppliers meet their ever-expanding expectations.

They have no patience with what was. If PC makers can't explain, easily and clearly, what their PC does—and didn't do before—they best forfeit the game and leave the playing field. Customers know what they want, how much they should pay for it, and what kind of response they should expect when something goes wrong.

Even giant IBM had to learn this lesson. It blew its early advantage in PCs by being oblivious to customers metamorphosing from tyros to tyrants, from naifs to sophisticates. In 1992, Big Blue woke up. It totally retooled its PC line to meet market demand. It dropped prices, lengthened warranties, and recruited platoons of IBM customer reps to trouble-shoot snafus over 24-hour hot lines.

But IBM is still playing catchup to customers' rising expectations. Smaller, nimbler market innovators have spoiled the customer. Dell Computer, a pioneer in the direct marketing of PCs, quelled rampant computer-phobia by offering customers brand-new user-friendly services; it initiated manufacturer-direct technical support, for example, shipping spare parts overnight and offering stellar phone support. Compaq Computer, meanwhile, cultivates direct relationships with information-technology managers through all sorts of third-party providers. QuickSource—an artificial intelligence program made by Inference Corporation—is shipped with Compaq's new network printer and holds the customer's hand, walking him or her safely through all the tricky turns. Nothing less will do.

For customers, the new competition is a boon. For competitors, it is a constantly sobering reminder of whose hands are now at the controls.

THE ROAD TO FAILURE

All of which brings us back to our original question: Why do some companies behave as if they were blind?

The main reason is that the threat they face looks so familiar. They've heard doomsayers and seen dark clouds massing many times before. Change, challenge, and crisis have all become cliches. A steady hand on the wheel, albeit guided by old navigation equipment, has previously helped them muddle through the storm. Why should they worry now?

Have you ever heard any of these rationalizations around your company?:

- "It's really not a very good value our competitor is offering, because it doesn't include a lot of our features." That's what American Airlines said about Southwest Airlines' policy of providing no meals, no luggage handling, and no long flights.
- "It's a good value but not in our preferred customer market." GM, Ford, and Chrysler ignored Toyota because it entered the less-desirable, low-end of the automobile market.
- "Sure they're hurting us, but with their unfair advantage, what can we do?" That's how U.S. Steel handed a large part of the United States market to Japanese producers.
- "The rules we're playing by have always worked before." That was American Express explaining why it could charge restaurateurs and merchants higher fees than other card issuers did—before AmEx was hit by the so-called Boston Revolution.
- "Those customers were a pain and not even profitable, so good riddance to them." That's how American television manufacturers first gave away the black-and-white market. We all know the rest of the story.

Eventually, of course, the signs of a failing value proposition cannot be ignored. Growth stalls, margins shrink, customers take flight. As the slide in profits turns into a free fall, management scrambles, resorting to a series of bolt-on tactics—temporary quick fixes—to address the most pressing problems and to bolster near-term performance. We've all witnessed these kind of initiatives:

- Selectively offer discounts to hold business that has started to go elsewhere.
- Introduce new promotions, terms, conditions, and offers to confuse and cloud the market.
- Beef up customer service by adding people to fix screwups and expedite delayed shipments.

■ Delay capital investments and adjust accounting methods to portray quarterly financial results more favorably.

■ Introduce "new and improved" products that are new in form, but not in substantive ways that are of consequence to purchasers.

These actions, all too familiar, are hollow and at times cynical efforts to veil a company's declining value to customers. The initiatives treat the symptoms, rather than the root causes of the performance malaise.

What is needed is a thorough renovation of the machinery that creates value—the operating model—so that management can reestablish leadership in the dimensions of value it has chosen to offer customers. Management teams, distracted by brush fires, too often can't see that their house is burning. They fail to take time for a thorough renovation to meet the new code of competition. Often, they don't possess the skill. The chief fire marshals in many companies come from a generation steeped in the techniques of operating a business, but with little experience in transforming one.

Also debilitating is that many management teams aren't teams at all. They're committees. In teams, every member focuses on a common goal—scoring a touchdown, dominating a market. In committees, members represent different functions or businesses and act to protect the interests of their spheres. This is the so-called stovepipe effect. If the management group is operating in a stable situation, a committee often works effectively. But during times of stress and change, committees tend to table core issues, because the resolution of those issues would necessarily change the balance of power and influence. Few managers tolerate having others wrestle away their coveted position in the pecking order.

So management committees, remembering the success of facile maneuverings in the past, resort to the familiar bolt-ons that brighten short-term performance. They fiddle with the fenders and chrome to get incremental adjustments to an operating model that works only in a steady-state world.

But in a market where the customers have taken control, where expectations of value are rising fast and today's leaders can be left by the wayside tomorrow, bolt-ons not only don't work, they actually make the situation worse. First, they drug the management team into a world of illusion, a world where they feel that they are solving the problems. The lost opportunity cost of these bolt-ons is enormous.

Second, bolt-ons add to the complexity of an already tired and underperforming operating model. When it comes time to revamp the model, the starting position is far more complicated. Basic problems have become enmeshed in a complex organizational snarl.

How many times have we heard the tales of executives ousted from companies overwhelmed by change? We hear the familiar stories of DEC and IBM, of Kodak and Westinghouse, of General Motors and Eli Lilly, of American Express and Steelcase. All are companies that have recently resorted to ritual public execution of their most senior executives: They fired their CEOs and are in the throes of a turnaround. Some will make it. Others most certainly will not.

What's needed in response to the new competition is focus and discipline—unprecedented focus and discipline—to define an unmatched value proposition, build an operating model, and sustain it through constant transformation and improvement.

2

THE NEW RULES OF COMPETITION

.

CHAPTER 2

THE NEW RULES OF COMPETITION

Is this any way to do business?

During the Gulf War in 1991, when the flow of Middle East crude oil was interrupted, gas prices rose 10 to 15 cents a gallon at most retail stations. But prices at the pumps of one major oil company stayed down. Its stations in Chicago charged 10 cents less for a gallon of regular gasoline than other brand-name retailers. By not hiking prices, the company shorted that year's revenues and profits by tens of millions of dollars.

Or how about this?

On the third floor of the headquarters of a large consumer electronics company, marketers are planning an extravagant launch for a new mini-camcorder. They expect the product to become the company's hottest-selling item. If they are correct, the mini-camcorder will generate the best profit margin the company has ever earned on a single product. At the same time, on the seventh, eighth, ninth, and twelfth floors, four other teams are competing among themselves to create a still better mini-camcorder with the goal of making the soon-to-be-unveiled one obsolete.

Or this?

A member of the sales staff at a home-improvement store phones a customer to ask if the new attic fan he bought last week is working all right. Just fine, the customer says. Good, replies the salesperson, who then asks if the customer has any other problems around the house that the store might help with. The dimmer switch in the dining room isn't working, the customer says. I'll drop a new one off tonight on my way home, the salesperson says, and I'll show you how to install it. Total sale: $6.71. The store's profit, allowing for the cost of the salesperson's time: nothing. A loss, in fact.

Do any of these stories suggest smart ways for a company to make money? In fact, they all do, and the companies—Atlantic Richfield Company (Arco), Sony, and Home Depot—are clear leaders in their markets. Just look at the numbers:

- Arco has over the past five years averaged a 20 percent return on equity, more than triple the average of its industry.
- Sony outperforms its competitors by almost every financial measure. Its average annual revenue growth over the past five years was 29 percent, compared to 14 percent for rivals. Sony's annual return on equity in the same period outpaced other consumer electronics companies.
- Home Depot has averaged 37 percent annual sales growth over the past five years, almost triple the average for the industry. Return on equity over the same period came to 26 percent, more than twice the industry average.

These companies aren't doing just a little better than the competition; they are leaving them in the dust. So if certain of their practices seem contrary to sound business doctrine, maybe it is the doctrine that needs looking at.

Let's consider the principles the examples above demonstrate. Because Arco produces its own crude oil on Alaska's North Slope, its production wasn't threatened by the Middle East conflict. It could have reaped a windfall, but by not raising prices Arco reaffirmed its commitment to its customers of being the low-price leader, and after announcing its price freeze, it enjoyed a 20 percent increase in sales overnight. What's more, by maintaining its low prices, Arco avoided the temptation to relax its stringent cost discipline.

Sony's first mini-camcorder became the most successful item in the home electronics market. Because of the product's success, however, Sony knew that a handier, lighter, perhaps more user-friendly product would soon appear. The company didn't want to see the Panasonic or Canon name on this product. So Sony rendered the first mini-camcorder obsolete before wringing every last penny of profit out of it. It gained in two ways: by boosting its market reputation as a product innovator and by improving its ability to innovate.

The Home Depot sales associate telephoned the customer who had bought the attic fan because the man lived in a neighborhood of expensive, older houses. That phone call probably earned Home Depot the loyalty of a customer who would make major home improvement and repair purchases over the next few years. The loss on the dimmer switch was a small down payment on a long, profitable relationship.

All three companies are hugely successful, and all three are selling something more than the obvious. This something more is customer value. Not just ordinary, but superior value. And not just superior value, but *continually improving* superior value.

Arco's superior value resides in its assurance to customers that, today and tomorrow, they won't find better gasoline prices elsewhere and needn't bother trying. With Sony, the superior value does not reside just in the camcorder, but also in the comfort customers can take from knowing that whatever product they buy from Sony will represent the state of the art. With Home Depot, the superior value resides in the consistently high level of helpful advice and service offered.

By delivering superior value, each of these three companies has changed its customers' expectations. In effect, these companies became market leaders not by fulfilling old-fashioned ideas of value, but by getting their business to master one band in the value spectrum. They don't try to be the best in everything. They believe in three important truths that characterize the new world of competition:

■ Different customers buy different kinds of value. You can't hope to be the best in all dimensions, so you choose your customers and narrow your value focus.

■ As value standards rise, so do customer expectations; so you can stay ahead only by moving ahead.

■ Producing an unmatched level of a particular value requires a superior operating model—a "machine"—dedicated to just that kind of value.

DIFFERENT CUSTOMERS BUY DIFFERENT KINDS OF VALUE

Let's define customer value more precisely; after all, the value provided by Arco, Sony, and Home Depot comes in very different forms. Customer value is the sum of benefits received minus the costs incurred by the customer in acquiring a product or service. Benefits build value to the extent that the product or service improves the customer's performance or experience. Costs include both the money spent on the purchase and maintenance, and the time spent on delays, errors, and effort. Both tangible and intangible costs reduce value.

Price, product quality, product features, service convenience, service reliability, expert advice, and support services can either create or destroy value for the customer. The value added or destroyed depends on how much the value exceeds or falls short of customer expectations.

As we began to study 80 market-leading companies, we noted that their customers tended to fall into a small number of categories. Some customers, including those of 3M and Nike, view a product's performance or uniqueness as the pivotal component of value. Price played some role in their decision-making because there is a limit to how much they will pay. But product results matter most.

A second group of customers, which include those of Nordstrom and Airborne Express, most value personalized service and advice. They can't be satisfied with standard products or fair prices. They want their individual requirements met. Companies that serve such customers emphasize relationships, develop an intimate knowledge of the customer's needs, and make a commitment to providing a total solution.

A third group of customers looks largely for the lowest total cost, through some combination of price and dependability. Customers of FedEx, Hertz #1 Club Gold, and McDonald's are examples. For them, no-hassle, speedy service is paramount. Customers of companies like Southwest Airlines, PriceCostco, and Arco, which put a high premium on price, fall into this group as well.

Value among the customers of these three categories of market-leading companies has come to mean three different things: best products, best total solution, best total cost. Market leaders choose to excel in delivering extraordinary levels of one particular value.

These companies have created a set of expectations in customers' minds that competitors now must strive to meet. They abide by the first of four new rules that govern market leaders' actions:

■ Rule 1: Provide the best offering in the marketplace by excelling in a specific dimension of value.

Market leaders first develop a value proposition, one that is compelling and unmatched. This rule does not mean that a company that focuses on price can ignore fashion or technological advances, or that it can deny its customers convenience. Any market leader, whatever value it chooses to deliver, must maintain reasonable standards in the other dimensions as well. But it doesn't have to excel in all of them—just one.

Our experience as consultants indicates that customers are able to distinguish among the various kinds of value, and they generally won't demand them all from the same supplier. Wal-Mart's proposition is "always the low price, always." Nobody goes there expecting personalized service. No one buys designer fashions from Bloomingdale's expecting a low price. Customers know that to expect superior value in every dimension from the same supplier is unreasonable.

■ Rule 2: Maintain threshold standards on other dimensions of value.

As Yugo found out, having the lowest-cost automobile on the market wasn't enough when the package also included subpar quality and service standards. As Apple and Compaq discovered, leadership in technology, innovation, or product performance wasn't enough when customers demanded lower prices. And as Nordstrom and Home Depot know, heaps of advice and warm, personal service can get a cold shoulder from customers if the companies fall short in attractive pricing or hassle-free basic service.

The rule is that you can't allow performance in other dimensions to slip so much that it impairs the attractiveness of your company's unmatched value. However, you don't have to strive to be the best on these other dimensions. Instead, channel energy into what separates you from the pack—and perform ably and adequately in other areas.

AS VALUE STANDARDS RISE, SO DO CUSTOMER EXPECTATIONS

Market leaders raise expectations and value norms not only in their own industries, but across the board. Customers are being conditioned—spoiled, some may say—to anticipate lower prices, speedier service, and more innovative products from all of their suppliers. So market leaders are finding that their preeminent position is always under siege from two directions: from rivals that focus on the same kind of value as they do and from other companies that increase customer expectations on the secondary dimensions of value.

Nike is not just facing threats from Reebok, another product performance leader. It's also facing Wal-Mart, which is changing customers' notions of what a pair of running shoes should cost. In turn, Wal-Mart can't relax when companies like PriceCostco are redefining the cost and convenience standards that Wal-Mart itself established. To sustain market leadership, it is not enough to deliver today's best product, price, or total solution; you must also be able to deliver tomorrow's and the next day's. To sustain its position at the top, a market leader must ensure that its operating model improves faster than the competition's—which leads to the third new rule:

■ Rule 3: Dominate your market by improving value year after year.

If one company could somehow deliver the absolute best in all dimensions of value, it would surely own the market. But no company can be the best at everything. When a company focuses all its assets, energies, and attention on delivering and improving one type of customer value, it can nearly always deliver better performance in that dimension than another company that divides its attention among more than one. Market leaders understand this as another truth in the new world of competition.

PRODUCING UNSURPASSED, EVER-IMPROVING VALUE REQUIRES A SUPERIOR, DEDICATED OPERATING MODEL

When we took a deep look at the market-leading companies, we invariably saw a wide variety of value propositions. We also saw the specific ways the companies structured their operations to deliver on their proposition—what we call operating models. Strikingly, the operating models were clustered not by industry, but by which value proposition—best product, best total cost, or best total solution—the company was pursuing.

In other words, the operating models of market-leaders pursuing the same value proposition in different industries are remarkably similar. Wal-Mart, FedEx, Schwab, Taco Bell, Southwest Airlines—all deliver lowest total cost and all focus on a similar combination of operating processes, management systems, business structure, and culture. Thus, an executive at one of these companies could move with relative ease across industry boundaries to another lowest-total-cost company, because the mechanisms for delivering value and making money are so similar.

Consider, for example, the gasoline retailing business. It illustrates how tightly linked the value proposition and operating model of a market leader are. Everyone knows that over the last decade the number of gasoline service stations has declined, that those remaining have become more price-competitive, and that the companies that operate them have gone to great lengths to increase the revenues, margins, and profits generated from their expensive corner real estate. They've added car washes, convenience stores, fast-food outlets, video rentals, and dry cleaning drop-off and delivery. Gas retailers have had no shortage of ideas.

Imagine that you are a strategic planner in the service station business. How might you make your business the most competitive in the market?

Well, you could automate the pumps. You could offer promotional giveaways on convenience items like milk and motor oil. You could take credit cards at the pump, or take other companies' credit cards. You could run specials on certain days of the week. You could give away soft drinks, glasses, or movie tickets. You could offer computer-generated maps. You could install a convenience store or a car wash. With a week to think about it, anyone could come up with 100 ideas for spurring sales. However, the competition could come up with the same list or an even better one. In fact, the competition doesn't even have to think about it. If your ideas are any good, they'll copy them. And if the ideas are no good, the competition will let you learn that lesson yourself, at your own expense.

In other words—and this applies to more than just the retail gasoline business—if you want, you can try to match every move made by your competitors and then up the ante a bit. It's a strategy lots of companies follow—"keeping up with the Joneses." But in the end, you'll offer no more than the next guy, and therefore be no better off.

On the other hand, you could decide to do one thing better than anyone else—to focus on price, for instance. Then you could direct all your creative energies to selling the cheapest gas in your market. That's what Arco did.

Arco built an integrated system, starting with its own North Slope crude oil and an efficient refinery tuned to refine only that grade of crude. It pulled its distribution system back to the West and Midwest in order to stay closer to its refinery. As a result, Arco now dominates California with a clear price advantage that customers recognize. It will perform well enough in other ways to avoid irritating customers, but it doesn't try to keep up with the Joneses. It's focused, specialized. Only if an idea lowers Arco's cost and thereby enhances its value proposition, will the company pursue it.

Could another company compete with Arco? Of course, but it would be difficult to do so on price. For argument's sake, let's assume that Arco has nailed down the price advantage. Some other company might decide to be the fastest gasoline retailer; every customer will get served, pay, and depart in 60 seconds. It could build its operating model around that proposition. Its dollar price might be a little higher than Arco's, but not so much higher that it would bother customers who value time.

How could a company do it? It could put its stations on easily accessible lots. It could spend R&D money on fast-pump technology. Its attendants could wear computerized credit-card terminals on their belts like the ones car rental companies use.

The point is that if the customer value you decide to offer is speed, the ideas automatically fall into place on how to create an operating system. The particular value that you decide to offer has the effect of defining your thinking about your business—of shaping the company's operating model. Low price, as we have seen, defines everything that Arco is and does. Once it chose to be the price leader in its market, the notion of installing computerized map machines or other gimmicks became irrelevant. The company is passionate about the one value it delivers to customers. Arco is behaving consistently with the fourth new rule for market leaders:

■ Rule 4: Build a well-tuned operating model dedicated to delivering unmatched value.

In a competitive marketplace, improving customer value is the market leader's imperative. The operating model is the key to raising and resetting customer expectations. Improving it can make competitors' offerings look less appealing, or even shatter their position by rendering their value proposition obsolete. The operating model is the market leader's ultimate weapon in its quest for market domination.

WHAT'S DIFFERENT

For some readers, the new rules of market leadership will be intuitively appealing. To others, they will raise stubborn questions. How can a company offer the best value proposition in the market (read: give its products and services away), and still make money? How can a company provide better customer value every year and still make money in the long run? Won't you burn out your employees if you're never satisfied with the level of value you offer—if you're on this treadmill of ever-expanding ambition? Isn't delivering value to customers in potential conflict with delivering value to shareholders?

We think not. In fact, in all market-leading companies we observed—corporations like Wal-Mart, Southwest Air, FedEx, Glaxo, Airborne, and Intel—customer value, shareholder wealth, and employee satisfaction move in lockstep. These companies view customer value as the indispensable source of both shareholder value and employee satisfaction. Without customer value, there is no sustainable business.

Some readers will ask what's new and different in our perspective. How does it relate to accepted wisdom that learning organizations, customer loyalty, and core competencies—to name a few of today's popular notions—contain the answers to all management woes? Our conclusion is that none of these notions gets to the heart of what sustains success in a competitive marketplace. At best, they provide partial solutions. Their relevance depends entirely on whether and how well they are channeled toward the pivotal issue of increasing customer value, year after year.

For example, the concept of core competence is that a company succeeds by leveraging what it's good at. Honda has a core competence in small engines. It has leveraged that capability in many markets, from motorcycles and autos to lawnmowers and generators. But Briggs and Stratton has a core competence in small engines, too. Why hasn't it been as successful? The answer lies not in examining these two companies' core competencies, but by understanding that they have different value disciplines. Honda, dedicated to the value discipline of best product, has built an operating model that naturally leverages its small engine competence into new application markets. Briggs and Stratton, with a value discipline focused on best total cost, has built an operating model that channels its small engine competence toward making its engines cheaper and cheaper.

Core competencies may be part of the operating model, but they aren't sufficient. They don't, like the operating model, help managers to balance the management of core and secondary processes, structure, and culture. 3M may have developed nonwoven technology as a core competency, but that's not enough to make it a product leader in tapes and soap pads. Likewise, suggesting that Wal-Mart's success stems solely from its logistics competency, or that Intel's success stems solely from its microprocessor design competency, pushes the concept of core competencies too far. Success is more multifaceted.

Similarly, the pursuit of customer satisfaction and loyalty doesn't by itself create unmatched value. Value comes from choosing customers and narrowing the operations focus to best serve those customers. Customer satisfaction and loyalty are simply the by-product of delivering on a compelling value proposition—not the drivers behind it.

Those companies that wish to sail at the head of their markets must weigh anchor from a mooring secured to value, namely the value proposition. That proposition must stress just one particular kind of value that customers want. Leaders will not pursue a diffused business strategy, but must continually focus on running a tight ship where their business practices enhance the one special value that they can provide better than anyone else.

3

THE WINNER'S CHOICE

CHAPTER 3

THE WINNER'S CHOICE

The sales people from FedEx no longer knock on the door at National Parts Depot. That's because Airborne Express, the Seattle-based $1.5-billion air express company, stole NPD's business away from FedEx with better service. Airborne has similarly pushed its competitors aside at Xerox, and Luxottica, the Long Island-based distributor of eyeglass frames.

Airborne's success at stealing business stems from many small strengths. But the big reason was its decision to deliver a different kind of value than FedEx. Whereas FedEx has chosen to create customer value through excellence of execution, Airborne has chosen to create value through excellence in customer care.

The juxtaposition of FedEx and Airborne highlights what particularly struck us in our study of 80 market leaders: In the same way that customers cluster into three different categories, as mentioned in the last chapter, companies cluster into distinctively different "value disciplines." These disciplines are based not on industry, but upon what kind of value proposition the companies pursued—best total cost, best product, or best total solution. We gave these three value disciplines, each appropriate for a different kind of customer, three distinctly different names: operational excellence, product leadership, and customer intimacy.

By operational excellence, we mean providing customers with reliable products or services at competitive prices, delivered with minimal difficulty or inconvenience. By product leadership, we mean providing products that continually redefine the state of the art. And by customer intimacy, we mean selling the customer a total solution, not just a product or service.

FedEx falls into the category of operational excellence, Airborne into the category of customer intimacy. Companies such as 3M, Nike, Motorola, and Sony fall into the category of product leadership. These companies have taken their leadership positions by narrowing their business focus, not broadening it. In line with the new rules of competition we set out in the last chapter, they chose a value proposition that highlighted a particular strength. With it, they developed a matching operating model to deliver that value. And they disciplined themselves to stick to and continually improve their combination of value proposition and operating model, while resisting the temptation to broaden their scope. When a company selects and pursues one of these value disciplines, it ceases to resemble its competitors.

THE OPERATING MODEL

The choice of a value discipline shapes the company's subsequent plans and decisions, coloring the whole organization, from its culture to its public stance. To choose a value discipline—and hence its underlying operating model—is to define the very nature of a company. What sets the inner workings of market leaders apart from their also-ran competitors is the sophistication and coherence of their operating models.

Operating models are made up of operating processes, business structure, management systems, and culture, all of which are synchronized to create a certain superior value. At the heart of the operating model sits not one but a set of core processes that make or break an organization's ability to create unsurpassed value at a profit.

Different value disciplines demand different operating processes. For instance, if your customers love your consistency and speed in delivering a value-for-the-money burger—as is the case with operationally excellent McDonald's—you'd better be stellar at the core processes of product supply, expedient customer service, and demand management. At the same time, you'll fine tune your structure to empower the people who can make a difference in producing value. You'll design your management systems around measuring and rewarding what's most important. And you'll make sure that your staff is indoctrinated with your specific definition of success.

In the other two disciplines, the operating model revolves around different core processes. If you're a product leader, such as Sony or Johnson & Johnson, the critical processes include invention, product development, and market exploitation. If you're a customer-intimate company—Home Depot, for example, or Cable & Wireless, the telecom corporation—you'll demonstrate superior aptitude in advisory services and relationship management.

Companies that excel in the same value discipline have remarkably similar operating models. Arco and McDonald's, for example, are strikingly similar because both pursue operational excellence. Likewise, the management systems, business structure, and culture of product leaders such as Sony and Johnson & Johnson look alike. But across two disciplines, the similarities end. Send people from Arco to Sony, and they will think they are on a different planet. Even within an industry, market leaders pursuing different value disciplines, such as Wal-Mart and Nordstrom, look completely different. Moreover, homogeneity exists only among *leaders* in the same value discipline; mediocre performers look pretty much like other mediocre performers in their own industries.

Let's look at each of the three value disciplines.

OPERATIONAL EXCELLENCE

Operationally excellent companies deliver a combination of quality, price, and ease of purchase that no one else in their market can match. They are not product or service innovators, nor do they cultivate one-to-one relationships with their customers. They execute extraordinarily well, and their proposition to customers is guaranteed low price and/or hassle-free service.

PriceCostco, the Kirkland, Washington and San Diego-based chain of warehouse "club" stores, doesn't provide a particularly rich selection of merchandise—only 3,500 items compared to the 50,000 or more found in competing stores. But as a customer, you don't have to spend much time deliberating over what brand of coffee or home appliance to select. PriceCostco saves you that hassle by choosing for you. The company's *Consumer Reports* mentality leads to rigorous evaluation of leading brands and shrewd purchasing of just the one brand in each category that represents the best value. To add excitement to the whole shopping experience—that is, to get the customer to come again and again—new items are constantly sprinkled into the assortment to build anticipation and a "value-of-the-week" atmosphere, while the on-premise bakery wafts a delicious smell of fresh bread and pastry.

Behind the scenes, PriceCostco follows an operating model in which it buys larger quantities and negotiates better prices to pass along to customers. It also carries only items that sell well. The company's information systems track product movement—and move it does! This data drives stocking decisions that optimize floor space usage. The place hums. It runs like a well-oiled machine and customers love it.

Dell Computer is another master of operational excellence. Dell has shown PC buyers that they do not have to sacrifice quality or state-of-the-art technology to buy personal computers easily and inexpensively. In the mid-1980s, while Compaq concentrated on making its PCs cheaper and faster than IBM's, college student Michael Dell saw a chance to outdo both companies by focusing not on the product but on the delivery system. Out of a dorm room in Austin, Texas, Dell burst onto the scene with a radically different and far more efficient model for operational excellence.

Dell realized that he could outperform PC computer dealers by cutting dealers out of the distribution process altogether. By selling to customers directly, building to order rather than to inventory, integrating his company's logistics with its suppliers', and creating a disciplined, extremely low-cost culture, Dell undercut Compaq and other PC makers in price while providing high quality products and services.

Yet another, less well-known example of operational excellence is GE's "white goods" business, which manufactures large household appliances. It has focused on operational excellence in serving the vast market of small, independent appliance retailers.

In the late 1980s, GE Appliances set out to transform itself into a low-cost, no-hassle supplier to dealers. It designed its Direct Connect program in pursuit of that objective. Direct Connect required that GE reengineer several of its operating processes, redesign its information systems, reconfigure its management systems, and create a new mindset among employees. As a result, the company has lowered dealers' net cost of appliances and simplified its business transactions.

Historically the appliance industry has endorsed the theory that a loaded dealer is a loyal dealer. If a dealer's warehouse was full of a manufacturer's product, went the argument, the dealer would be committed to that company's product line because no room remained to stock goods from anyone else. Manufacturers' programs and pricing were built around the idea that dealers got the best price when they bought a full truckload of appliances and offered the best floor plan.

But changes in retailing caused GE to question that assumption. For one, the loaded-dealer concept was costly for independent appliance dealers, whose very existence was threatened by the growing clout of low-price, multibrand chains like Circuit City. Independent stores could hardly afford to match the large stock of the chains. Moreover, the chains could put price pressure on manufacturers, causing makers' margins to shrink.

Realizing that it had to supply high-quality products at competitive prices with little hassle, General Electric abandoned the loaded-dealer concept and reinvented its operating model—the way it made, sold, and distributed appliances. Under Direct Connect, retailers no longer maintain their own inventories of major appliances. They rely instead on General Electric's "virtual inventory," a computer-based logistics system that allows stores to operate as though they have hundreds of ranges and refrigerators in the back room when, in fact, they have none at all.

With Direct Connect, retailers acquire a computer package that gives them instant access to GE's on-line order-processing system 24 hours a day. They can use the system to check on model availability and to place orders for next-day delivery. The dealers get GE's best price, regardless of order size. Direct Connect dealers also get, among other benefits, priority over other dealers in delivery scheduling, plus consumer financing through GE Credit with the first 90 days free of interest. In exchange, Direct Connect dealers make several commitments: to sell nine major GE product categories while stocking only carryout products, such as microwave ovens and air conditioners; to ensure that GE products generate 50 percent of sales and to open their books for review; and to pay GE through electronic funds transfer on the 25th of the month after purchase.

Under the Direct Connect system, dealers have had to give up some float time in payables, the comfort of having their own back-room inventory, and some independence from the supplier. In return, they get GE's best price while eliminating the hassle and cost of maintaining inventory and assembling full-truckload orders. The result: Their profit margins on GE products have soared.

Virtual inventory, it turns out, works better than real inventory for both dealers and customers. "Instead of telling a customer I have two units on order," says one dealer, "I can now say that we have 2,500 in our warehouse. I can also tell a customer when a model is scheduled for production and when it will be shipped. If the schedule doesn't suit the customer, the GE terminal will identify other available models and compare their features with competitive units."

Meanwhile, GE gets half the dealer's business and saves about 12 percent of distribution and marketing costs. And since dealers serve themselves through the network, GE saves time and labor in responding to inquiries and in order entry; in fact, the Direct Connect system *is* the order-entry process. Most important, GE has gained a valuable commodity from its dealers: data on the actual movement of its products. Most appliance manufacturers have been unable to track consumer sales accurately because they can't tell whether dealers' orders represent requests for additional inventory or actual customer purchases. With Direct Connect, GE knows that vendors' orders are actual sales to customers.

GE links its order-processing system to other systems involved in forecasting demand and planning production and distribution. The company now, in effect, manufactures in response to customer demand instead of to inventory. It has reduced and simplified a complex and expensive warehousing and distribution system down to 10 strategically located warehouses that can deliver appliances to 90 percent of the country within 24 hours.

Businesses like PriceCostco, Dell Computer, and GE Appliances, which have vigorously pursued a strategy of operational excellence, have built an operating model based on four distinct features:

■ Processes for end-to-end product supply and basic service that are optimized and streamlined to minimize costs and hassle.

■ Operations that are standardized, simplified, tightly controlled, and centrally planned, leaving few decisions to the discretion of rank-and-file employees.

- Management systems that focus on integrated, reliable, high-speed transactions and compliance to norms.
- A culture that abhors waste and rewards efficiency.

PRODUCT LEADERSHIP

A company pursuing product leadership continually pushes its products into the realm of the unknown, the untried, or the highly desirable. Its practitioners concentrate on offering customers products or services that expand existing performance boundaries. A product leader's proposition to customers is best product, period.

A product leader consistently strives to provide its market with leading-edge products or useful new applications of existing products or services. Reaching that goal requires that they challenge themselves in three ways. First, they must be creative. More than anything else, being creative means recognizing and embracing ideas that may originate anywhere—inside the company or out. Second, they must commercialize their ideas quickly. To do so, all their business and management processes are engineered for speed. Third and most important, they must relentlessly pursue ways to leapfrog their own latest product or service. If anyone is going to render their technology obsolete, they prefer to do it themselves. Product leaders do not stop for self-congratulation; they are too busy raising the bar.

Johnson & Johnson meets all three of these challenges. It brings in new ideas, develops them quickly, and then looks for ways to improve them. In 1983, the president of J&J's Vistakon, Inc., a maker of specialty contact lenses, heard about a Copenhagen ophthalmologist who had conceived a way of manufacturing disposable contact lenses inexpensively. At the time, Vistakon generated only $20 million in annual sales, primarily from a single product, a contact lens for people with astigmatism.

Vistakon's president got his tip by telephone from a J&J employee who worked for Janssen Pharmaceutica, a Belgian drug subsidiary. Instead of dismissing the ophthalmologist as a mere tinkerer, these two executives speedily bought the rights to the technology, assembled a management team to oversee development, and built a state-of-the-art facility in Florida to manufacture disposable contact lenses called Acuvue.

By the summer of 1987, Acuvue was ready for test marketing. In less than a year, Vistakon rolled out the product across the United States with a high-visibility ad campaign. Vistakon—and its parent, J&J—were willing to incur high manufacturing and inventory costs before a single lens was sold. Vistakon's high-speed production facility helped give the company a six-month head start over would-be rivals such as Bausch & Lomb and Ciba-Geigy. Caught off guard, the competition never caught up. Vistakon also took advantage of the benefits of decentralization—autonomous management, speed, and flexibility—without having to give up the resources, financial and otherwise, that only a giant corporation could provide.

In 1991, Vistakon's sales topped $225 million worldwide, and it had captured a 25 percent share of the U.S. contact lens market. Part of the success resulted from directing much of the marketing effort to eye-care professionals to explain how they would profit if they prescribed the new lenses. In other words, Vistakon did not market just to consumers. It said, in effect, that it's not enough to come up with a new product; you have to come up with a new way to go to market as well.

J&J, like other product leaders, works hard at developing an open-mindedness to new ideas. Vistakon continues to investigate new materials that would extend the wearability of the contact lenses and even some technologies that would make the lenses obsolete. Product leaders create and maintain an environment that encourages employees to bring ideas into the company and, just as important, to listen to and consider these ideas, however unconventional. Where others see glitches in their marketing plans or threats to their product lines, companies that focus on product leadership see opportunity and rush to capitalize on it.

Product leaders avoid bureaucracy at all costs because it slows commercialization of their ideas. Managers make decisions quickly since, in a product leadership company, it is often better to make a wrong decision and correct it than to make a decision too late or not at all. That is why these companies are prepared to decide today, then implement tomorrow. Moreover, they continually look for new ways—such as concurrent engineering—to shorten their cycle times. Japanese companies, for example, succeed in automobile innovation because they use concurrent development processes to reduce time to market. They do not have to aim better than competitors to score more hits on the target because they can take more shots from a closer distance.

Companies excelling in product leadership do not plan for every possible contingency, nor do they spend much time on up front detailed analysis. Their strength lies in reacting to situations as they occur. Fast reaction times are an advantage when dealing with the unknown. Vistakon's managers, for example, were quick to order changes to the Acuvue marketing program when early market tests were not as successful as they had expected. They also responded quickly when competitors challenged the safety of the lenses. They distributed data combating the charges, via FedEx, to some 17,000 eye-care professionals. Vistakon's speedy response engendered goodwill in the marketplace.

Product leaders have a vested interest in protecting the entrepreneurial environment that they have created. To that end, they hire, recruit, and train employees in their own mold. When it is time for Vistakon to hire new salespeople, for example, its managers do not look for people experienced in selling contact lenses; they look for people who will fit in with J&J's culture. That means their first question isn't about a candidate's related experience; it's more likely to be, "Could you work cooperatively in teams?" or "How open are you to criticism?"

Product leaders are their own fiercest competitors. They no sooner cross one frontier, than they are scouting out the next. They have to be adept at rendering obsolete the products and services that they have created. They realize that if they don't develop a successor, another company will. J&J and other innovators are willing to take the long view of profitability, recognizing that extracting the full profit potential from an existing product or service is less important than maintaining product leadership and momentum. These companies are never blinded by their own successes.

Not surprisingly, the operating model of the product leader is very different from that of the operationally excellent company. Its main features include:

- A focus on the core processes of invention, product development, and market exploitation.
- A business structure that is loosely knit, ad hoc, and ever-changing to adjust to the entrepreneurial initiatives and redirections that characterize working in unexplored territory.
- Management systems that are results-driven, that measure and reward new product success, and that don't punish the experimentation needed to get there.

■ A culture that encourages individual imagination, accomplishment, out-of-the-box thinking, and a mind-set driven by the desire to create the future.

CUSTOMER INTIMACY

A company that delivers value via customer intimacy builds bonds with customers like those between good neighbors. Customer-intimate companies don't deliver what the market wants, but what a specific customer wants. The customer-intimate company makes a business of knowing the people it sells to and the products and services they need. It continually tailors its products and services, and does so at reasonable prices. Its proposition is: "We take care of you and all your needs," or "We get you the best total solution." The customer-intimate company's greatest asset is, not surprisingly, its customers' loyalty.

Customers don't have to be resold through expensive advertising and promotion. Customer-intimate companies don't pursue transactions; they cultivate relationships. They are adept at giving the customer more than he or she expects. By constantly upgrading their offerings, customer-intimate companies stay ahead of their customers' rising expectations—expectations that, by the way, they themselves create. Home Depot is a good example of a company that is better than most at building relationships that pay off in repeat sales from a loyal customer base.

However, the high-water mark for customer intimacy probably was set by IBM in the 1960s and 1970s. Customers never looked to IBM for the hottest product. In fact, IBM's response to customers who asked about leading-edge technology was always, "Just wait 18 months, and we will have that, too." It was not that IBM didn't invest in product innovation, but it knew that product innovation was not the central value proposition binding customers to the company. Best price wasn't part of the company's proposition, either. That was left to the plug-compatible computer makers, such as Amdahl.

So if IBM didn't mean best price or best technology, what did it mean? IBM was a comfort, a friend. IBM's people knew the heads of data processing, knew what their problems were, knew how to help them solve those problems and look good to their bosses. IBM assisted them with applications planning and technology architecture. It helped them fight for budgets and, through its executive education programs, get their bosses to appreciate technology. IBM's central value proposition was delivering a total solution in a customer-intimate fashion.

Customer-intimate companies consider the customer's lifetime value, not just the profit and loss on a few transactions. Their employees make sure that each customer gets exactly what he or she really wants. These companies have designed operating models that allow them to produce and deliver a much broader and deeper level of support. They tailor their mix of services or customize the products, even if it means acting as a broker to obtain these services and products from third parties or co-providers.

Cable & Wireless Communications, based in Vienna, Virginia, has worked for years to become a customer-intimate organization. It is the world's largest long-distance company devoted entirely to business customers. Cable & Wireless attributes its 20 percent annual growth rate in the number of long-distance customer minutes to its striving continuously to serve its customers better than its bigger competitors—such as MCI.

Cable & Wireless executives knew long ago that their long-distance operation couldn't compete on price with the Big Three, AT&T, MCI, and Sprint. So they sought to differentiate themselves by providing the best ongoing customer support in the industry, along with direct sales consultation that gives the sales force an intimate knowledge of what makes its customers successful. The result is that Cable & Wireless has turned itself from a mundane commodity business peddling long-distance service into a sophisticated telemanager, a partner with its customers. Does the customer need 800 service that routes calls, blocks calls, or captures data? Cable & Wireless supplies the expertise and information systems. "The product is conceived at the customer's office," says president and chief operating officer Gabriel Battista.

Cable & Wireless pins its success on choosing the customers it can serve best—small to medium-size businesses with monthly billings of $500 to $15,000. In such small businesses, Cable & Wireless's 500 U.S. salespeople, working out of 36 regional offices, can actually act like telecommunications managers. Corporations too small to hire their own telecom gurus value the advice and expertise Cable & Wireless people can offer.

Cable & Wireless then goes on to segment its small to medium-size business market vertically. By refining its market segments, it can appeal to specific customers with specialized services that no other company can begin to provide. One of its customer segments is the legal profession. Cable & Wireless is developing features and functions that have tremendous appeal to lawyers, such as innovative ways to track and segment billing of calls linked to specific client accounts. "We want to sell products that fit the legal industry like a glove," says Battista.

Cable & Wireless then takes the next step and fine tunes its services to each customer. If that means something as simple as printing its bills on both sides of the paper, Cable & Wireless obliges. The company wants customers to feel they're getting the support of not just the sales force but of the entire company.

Cable & Wireless empowers all employees who work with customers to make the most sophisticated decisions possible. Pricing was once the domain of corporate pricing gurus. No longer. Each of the 50 local managers has his or her own pool of funds to structure pricing. The same thing goes for promotional, advertising, and trade-show money. The corporate center doesn't hog the budget and issue edicts. The local managers allocate money as they see fit. They prepare budgets and send them up the corporate ladder.

Do Cable & Wireless managers run amok with so much authority? It can happen, Cable & Wireless executives concede. But if so, they figure that the occasional screwup is worth it. Executives go on to audit all decisions and practices to both catch blunders and help the front lines learn from them.

All of these practices help Cable & Wireless people build very tight relationships with customers. The result is extremely high customer retention rates: Cable & Wireless loses only 2 percent of long-distance minutes billed each month, compared to an industry standard of 3 percent to 5 percent.

Of course, only through those high retention rates can the company continue to fund its high level of support. One hundred percent of the sales force is dedicated to the dual objectives of providing ongoing help to customers and bringing in new customers. For large accounts, the company also assigns strategic support representatives, to whom the customer can turn at any moment for hand-holding. Cable &

Wireless also has what the company calls a "retention day," during which salespeople will sit down with large accounts to go over every aspect of service.

The company holds out a big carrot to keep its people focused on customer retention—it compensates its sales force based on how long a customer remains with the company. In addition, unlike competitors that pay salespeople according to the number of accounts landed and the dollars billed, Cable & Wireless compensates people based on their ability to retain existing accounts. Salespeople don't hesitate to suggest that customers switch to more appropriate services, even if the new services bring in less money. The result once again: happier, more loyal customers.

To move quickly in responding to customers, Cable & Wireless maintains state-of-the-art software capabilities, both to customize services such as billing, and to design and assemble its own switches. The company also operates an integrated information system so that, with a few keystrokes, anyone can bring up all pertinent information on a customer, from orders to billing. Cable & Wireless has worked hard in recent years to reengineer its processes to assure greater customer intimacy than any competitor can provide. "Ultimately," says Battista, "we see our competitive edge as our ability to look at our customers' needs and to customize our products and services to fit these needs exactly and uniquely, so they can reduce operating expenses, increase their competitive position, or become more productive."

Again, the operating model of the customer-intimate company is very different from that of businesses pursuing other disciplines. Its features include:

■ An obsession with the core processes of solution development (i.e., helping the customer understand exactly what's needed), results management (i.e., ensuring the solution gets implemented properly), and relationship management.

■ A business structure that delegates decision-making to employees who are close to the customer.

■ Management systems that are geared toward creating results for carefully selected and nurtured clients.

■ A culture that embraces specific rather than general solutions and thrives on deep and lasting client relationships.

WHY CHOOSE?

Choosing a value discipline is a fateful event in that it not only commits a company to a single path to achieve greatness, it also purposely destines the company to choose a secondary role in the other disciplines. That's because each discipline requires a company to emphasize different processes, to create different business structures, and to gear management systems differently. For example, when it comes to business structure, the product leader thrives on ad hoc and fluid structure to foster invention and allow resources to be redeployed quickly. Operationally excellent companies, on the other hand, do best with the major brain-trust at central locations where standard operating procedures get refined and decisions are made about acquiring and using capital-intensive assets. A natural organizational structure for the customer-intimate company is to move more of the decision-making responsibility out to the boundaries of the organization, closer to the customer.

Despite the specialization required of market leaders, we regularly come across managers who don't buy the idea of having to narrow their operational focus: "What you're saying about making hard choices doesn't apply to us," they say. "We're good at all three disciplines."

Yet when we look at these managers' businesses, we invariably find companies that don't excel, but are merely mediocre on the three disciplines. Sure, as the ante has risen in their markets, they've improved their cost structure and become more aware of their customers. They've added new products and line extensions over the years. They've kept up with rising parity levels to stay in the game. What they haven't done is create a breakthrough on any one dimension to reach new heights of performance. They have not traveled past competence to reach excellence. To these managers we say that if you decide to play an average game, to dabble in all areas, don't expect to become a market leader.

Thus, choosing a discipline is the choice of winners.

Not choosing means ending up in a muddle. It means hybrid operating models that are neither here nor there, and that consequently cause confusion, tension, and loss of energy. It means steering a rudderless ship, with no clear way to resolve conflicts or set priorities. Not choosing means setting yourself up to be overtaken by another player that is committed to unmatched value and focused on how to achieve it. Not choosing means letting circumstances control your own destiny. Not choosing means creating managerial complexity that results in your doing business with yourself, rather than with your customers. And that's exactly what will set you adrift in the stormy new seas of competition.

4

THE DISCIPLINE
OF OPERATIONAL
EXCELLENCE

CHAPTER 4

THE DISCIPLINE OF OPERATIONAL EXCELLENCE

Henry Ford knew about operational excellence. In fact, he practically invented it. The motor mogul built his manufacturing empire around a single notion—efficient production—and infused his whole company with that idea.

Today we would call the early Ford Motor Company a paragon of operational excellence, because the founder's business model was tuned to a single purpose: delivering an acceptable product at the lowest possible price. As Ford's costs fell, the retail price of the Model T car fell too, from $850 to $290.

Henry Ford's singular focus on achieving efficiency is the same idea that drives operationally excellent market leaders today. These companies—like Wal-Mart and Southwest Airlines—wave one bright banner high above the teeming marketplace: the promise of lowest total cost.

Lowest total cost? It *can* mean lowest price, but it doesn't always. What it does mean is that when all the costs to the customer of owning and using the company's product or service are added up—costs such as price, time spent at the checkout counter, the inconvenience of untimely repair—nobody else's deal is likely to be any better.

Some of today's operationally excellent companies could teach Henry Ford a thing or two. That's because while Ford focused solely on selling at the lowest price, many operationally excellent companies today focus on multiple tangible and intangible costs. To be sure, price remains the focus of most operationally excellent companies—prices so low that customers sometimes marvel: "How do they do it?" Wal-Mart and PriceCostco, for instance, continually surprise customers with prices their competitors wouldn't dream of offering. Service companies like Southwest Air and AT&T Universal Card Services similarly prompt customers to wonder, "Why don't they keep some of that money in their own pockets instead of giving it to us?"

When operationally excellent companies talk of low—or lowest—prices, they mean prices that are consistently low. Anybody can hold a fall clearance sale, an anniversary promotion, or a Presidents' Day extravaganza. Operationally excellent companies trumpet their low prices every day, 365 days a year.

When operationally excellent companies boast of their lowest *total cost*, they may, however, be emphasizing product reliability and durability, which lower customers' future costs of ownership. Toyota ads, for instance, show its products running on and on—for 200,000 miles, 300,000 miles, and more. Maytag touts its Rip Van Winkle repairman, whose sleep has been undisturbed for years. Timex used to boast its watches could "take a licking and keep on ticking." These companies' customers cherish the dependability they get along with the low price. The prices they paid look lower and lower as the trouble-free years roll on.

Another element of cost that operationally excellent companies stress is convenience—the absence of tangible or intangible costs stemming from annoyance and irritation. The strength of these companies lies in the delivery of swift, dependable service—the kind you get from, for instance, 800-Flowers, which accepts telephone orders from anywhere to ship flowers anywhere. It couldn't be easier or involve less total cost.

And Saturn Corp.—which may be closer to Henry Ford's idea of a car company than today's Ford—has brought the lowest total cost idea into its showrooms by eliminating one of the chief costs of buying a new car: the confrontation with the salesperson. Furthermore, Saturn dealers' service-delivery system makes the shop visit almost a pleasure.

Transactions that are easy, pleasant, quick, accurate—market-leading operationally excellent service companies like Charles Schwab and AT&T Universal Card design the means to achieve that end. Occasional mistakes happen, of course, but operationally excellent market leaders make sure they're so uncommon as to be remarkable. And when mistakes do happen, most recover with such panache that customers are left even more impressed than they would have been had the foul-ups never occurred.

But no matter what their formula for combining price, reliability, and hassle-free service to deliver lowest total cost, operationally excellent companies deploy an operating model based on a set of design principles handed down from Henry Ford. Ford's business was highly regimented, proceduralized, rule-driven. There was only one way—the efficient way—to do everything. Complex work was divided into simpler repetitive tasks and combined, via the assembly line, into an integrated process. The result: efficiency of effort *and* efficiency of coordination. Current thinking on business reengineering owes a debt to Ford. He built low-cost, no-frills factories. He aggressively pursued automation to minimize labor and to lower variable costs. The result was that he could design manufacturing processes and work procedures that demolished former standards of cost and performance.

Today, standardized assets and efficient operating procedures are the backbone of every operationally excellent company. It's not by accident that all Wal-Mart stores look alike, that all Southwest Airlines jets are similarly configured 737s, that all J.B. Hunt long-haul trucks are identical, and that all Taco Bell restaurants are as alike as their tacos. Every operationally excellent company that operates over a wide geography has built a network of no-frills, standardized assets that form the basis for efficient operating procedures.

But achieving and sustaining operational excellence requires more than cloning hyper-efficient assets. Today, as in Henry Ford's era, variety kills efficiency. Ford maintained a very narrow product line. He didn't introduce a variant of the Model T until millions of units of the basic model had been produced. As for variety in color, he left posterity his legendary remark: "Any color you want as long as it's black." Operationally excellent companies reject variety, because it burdens the business with cost. They produce no-frills products for the middle of the market where demand is huge and customers are more interested in cost than in choice.

Undisciplined companies, on the other hand, let products and services proliferate. They create a variant in response to one customer or operational demand, then create another to fill a different niche.

Since they can't be all things to all customers, operationally excellent companies work at shaping their customers' expectations. If price is their strong point, price is what they stress, and they make virtues of their apparent limitations. Schwab for example, crows about not having its own investment research and advisory services. Cut out the "biased" research and pocket the savings, the company tells customers. PriceCostco's product selection is slim, but its prices on category leaders are unbelievably low. Southwest Airlines doesn't offer meals, baggage handling, or advanced seating, but it lures short-haul business travelers with its frequent departure schedule and super-low prices.

Henry Ford maintained rigid, centralized control as his business grew because he was determined to capture the benefits of standardized procedures and economies of scale. His control system was based on detailed measures of every element of his operation. Ford sweated the details, and his management systems reflected it.

Today, Ford's principles of operational excellence are applied in industries as diverse as retail, brokerage, transportation, credit cards, and of course, manufacturing. Wal-Mart, for example, has built a supply process that integrates product flow from the supplier's factory all the way through to the store shelf. Wal-Mart, Charles Schwab, Southwest Airlines, FedEx, Taco Bell, and AT&T—all icons of operational excellence—know their activity-based costs and their transaction profitability. Their discipline is evident in their value propositions to the customer and in their operating models.

If Henry Ford were to run an operationally excellent company today, he would have to update his principles somewhat. For all the pluses of his operating model, he would now have to take into account today's enlightened employees, efficient transactions, information technology, and service intensive work. Let's look at each area in turn.

THE MANAGEMENT OF PEOPLE

Operationally excellent companies run themselves like the Marine Corps: The team is what counts, not the individual. Everybody knows the battle plan and the rule book, and when the buzzer sounds, everyone knows exactly what he or she has to do.

The heroes in this kind of an organization are the people who fit in, who came up through the ranks. They're dependable, like the FedEx driver who delivered the Denver-bound package before 10 A.M. in spite of the snowstorm that forced the plane off the runway and the truck into a drift. For the operationally excellent company, a promise is a promise. For the company's employees, dedication is paramount.

These companies aren't looking for free spirits. They want people who are trainable. They'll hire them and teach them the Wal-Mart—or FedEx, UPS, or Southwest—way of business. At McDonald's, the store manager knows and takes pride in the fact that the company president rose up through the ranks. He started, just as the manager did, flipping burgers and scooping fries. At McDonald's, as in most operationally excellent companies, what's important is not who you are but what the company will make out of you.

The employee of the year? That would be the best team player, who will get his or her name added to the plaque in the employee lunchroom. Peer recognition is the best compliment one can get and the plaque itself—inexpensive, inclusive, and public—symbolizes much about the company's culture. Avoiding waste is what the operationally excellent company is all about.

When Michael Dell visited the headquarters of Compaq, his principal competitor in the PC market, he walked around and looked at all the marble in the building lobby, all the rich furnishings. Then he went home and told his staff, "We can beat these guys." One look at the Dell headquarters and you know this PC company hasn't wasted any of its capital on office space. Similarly, Nucor Steel takes pride in the thrift signaled by its having sited its corporate headquarters in a Charlotte, North Carolina strip mall.

The point is that people at operationally excellent companies don't feel deprived by eating at restaurants without white linen tablecloths. They don't expect expense accounts to include a night at the Ritz—Motel 6 is more their style. They disapprove of ostentation. That's why operationally excellent companies can reward their employees not with big cash awards or broad empowerment, but by putting people in the limelight—often accomplished with an instant photo glued to that cheap plaque. The visible pat on the back goes a long way toward cultivating a workforce that is highly motivated and dedicated to the value proposition of the organization.

EFFICIENT TRANSACTIONS

Since the time of Henry Ford, the impact of technology has been immense. Information technology, for example, has automated routine tasks and coordinated activity through better communications. The impact on manufacturing productivity has been awesome—to the point that in many industries transaction costs, administrative expenses, and overhead have dwarfed production costs. As a result, every operationally excellent company strives for low overhead, with efficient, reengineered business processes. A revolution in transaction efficiency has swept nearly all businesses. This has caused a major revision of one of Henry Ford's guiding design principles.

The founding Ford believed deeply in building tightly integrated processes for the production of automobiles. He linked auto assembly to subassembly manufacturing, which was linked to component manufacturing, which in turn was linked to materials production. Eventually, this logic led Ford to almost complete vertical integration—steel mills, glass factories, even rubber plantations in Brazil—all to create a single-threaded manufacturing process. Ford used his management hierarchy to coordinate all of these activities, which coincidentally lowered transactions costs between activities.

More recently, however, companies have found ways to achieve even greater efficiency—not by vertically integrating, but by *virtually* integrating. Today, operationally excellent companies view themselves and their suppliers and distributors not as discrete, allied entities, but as members of a single product supply team. Streamlining the connections among team members eliminates duplications, delays, and even payment complications that come from arms-length handoffs. Customers consult with suppliers to ask: Why do we both inspect product quality—outgoing in your plant and incoming in ours? Why do my accounts payable people and your accounts receivable people duplicate each other's work? Why do I spend so much money on distribution centers, trucks, logistics, and warehousing capabilities, when you have them as well? Can't we cut those expenses in half? Why don't we figure out a way to make product flow freely, with maximum efficiency, from your company to ours?

Supplier Procter & Gamble and retailer Wal-Mart asked one another those questions, and then they turned purchasing and supply tradition on its head. Now product flows from P&G to Wal-Mart more smoothly than between internal departments in many companies. The agreement Wal-Mart made with P&G set new standards for every retailer's logistics arrangement with its suppliers.

Like Wal-Mart, other operationally excellent companies have redesigned the transaction process between themselves and their suppliers. Purchase authorizations, purchase orders, and backorder notices have become nearly extinct in new, continuous-replenishment processes. Bills of lading, receipt notifications, and shipping manifests, which add unnecessary costs to the flow of goods between companies, have disappeared, along with paper invoices and many of the clerks who processed them.

This continuous-replenishment concept is a simple one: Suppliers assume the responsibility for managing customer inventories, which, in return, allows them to smooth the flow of goods and lower their own end-to-end costs. Everyone wins. The operationally excellent company purchases lower-cost products and unburdens itself of much unnecessary work.

Wal-Mart, a pioneer in creating these cost-cutting relationships, today uses an electronic data-interchange system to send daily sales data to suppliers. The suppliers' computers integrate this data with, for instance, warehouse inventory information and sales forecasts to generate a new order, if one is necessary.

It's not just paper and its processing that such new arrangements cut. They slash distribution and transportation costs as well. The goal of integrated logistics, as the concept is broadly called, is to move product from maker to user in a single step. It treats the outbound logistics system of the supplier and the inbound logistics system of the customer as a single, integrated system. No time or motion is wasted—and no money either—on moving products in and out of intermediate warehouses. The philosophy: Product that isn't moving isn't being distributed.

Again, Wal-Mart offers an admirable example. It has developed a system of cross-docking in which two sets of trucks—one coming from supplier factories and the other set heading for stores—arrive simultaneously at a company loading dock. Workers move the product from the first set of trucks into the second, avoiding lengthly warehouse storage altogether. The product then heads to its final destination with no costly rest stops along the way. This concept is known as flow-through or one-stop logistics.

Tupperware Home Parties has similarly revamped its logistics operation, eliminating the distributors who warehoused inventories for dealers, who in turn ordered and picked up product for Tupperware party hosts, who only then delivered customer orders taken weeks earlier. With today's integrated logistics, dealers place orders by dialing with modem-equipped PCs into Tupperware's mainframe computer. Factories then ship via UPS directly to the dealer, party host, or customer.

The last step in the buying process between customer and supplier is billing and payment, and here, too, operationally excellent companies are exploring a new notion: invoiceless payment. In other words, they're eliminating the bill.

Again, let's look at Wal-Mart. When a customer buys Procter & Gamble disposable diapers at a Wal-Mart store, the checkout scanner records the sale and orders a credit sent to P&G. There's no need for intermediate steps.

Ford uses a similar process. When a supplier shipment arrives at a Ford assembly plant, no one at the dock checks to make sure the freight matches the invoice, because there is no invoice. Instead, a receiving clerk at the dock checks a computerized database of purchase orders issued by Ford. If the shipment corresponds to an outstanding purchase order, the clerk accepts the shipment and enters a confirmation in the database. The computer automatically cuts a check, which is sent to the supplier. Payment authorization takes place at the dock, not in accounts payable. As a result, Ford has been able to reduce by 20 percent the number of clerks in its accounts payable department.

As remarkable as these achievements might seem if viewed from, say, five years back, today's leading-edge practitioners of operational excellence have moved way beyond just reducing transaction costs. Imagine, for instance, that you're in the cereal business, and you get a call like this one from a major nationwide retailer: "About this cereal of yours," the caller says, "we've got some questions for you. We've examined the box and done a cost analysis of the contents. It looks to us like we're paying $3 for about 20 to 30 cents worth of product. Packaging might add another dime, which makes a total of 40 cents. We're wondering where the $2.60 difference is going? We don't understand. So we'd like to sit down with you to see if we can figure out how to kill the monster in your operation that's swallowing so much cash."

That kind of discussion is going on throughout the business world. Operationally excellent customers are invading their suppliers' domains.

McDonald's, Ford, and Wal-Mart, among others, are bent on achieving efficiency throughout the entire product supply process—even when it means stepping into someone else's fiefdom. If you do business with one of these companies, you'll achieve operational excellence in your own corporation—with their help—or you won't get their business.

What makes operational excellence necessary is the new competition. What makes it *possible* are the new computer information systems and networks, which play a vital enabling role in the creation of operationally excellent processes. Companies exploit today's low-cost, high-performance technology to increase coordination and control over their entire system and to speed and streamline individual tasks. Information systems have become not only the nervous system but also the backbone of their operations.

INFORMATION TECHNOLOGY

Because technology is so important in operationally excellent companies, one usually has to look inside the companies' computer systems to understand their core business processes. The systems—and related databases and applications—are so highly automated that they don't just track the process, they *contain* and *perform* it. For example, at FedEx anyone who deals with the movement of a package—driver, sorter, customer-service agent—uses the same package record to coordinate the work. At Hertz, the service agent at the airport counter, the technician readying a car for rental, and the agent who checks the car back into the parking lot all enter and extract data from the same system.

The power of information technology is especially evident in industries like securities brokerage. Charles Schwab has brought to that business an entirely different operating model, built on a sophisticated base of information systems. As a result, Schwab's cost structure is so much lower than Merrill Lynch's or Smith Barney's that it can make high margins while charging less than half the price for stock transactions. Further, because its systems communicate buy and sell orders directly to the floor of the stock exchange, they can execute trades and confirm prices while the customer is still on the phone. Traditional brokers are probably still writing your orders down on slips of paper. They'll have to get back to you.

It's not that Merrill Lynch and Smith Barney don't want systems like Charles Schwab's. Goodness knows they've spent enough money trying to build them. It's just that these companies don't adapt well to the organizational demands that information technology makes. Without organizational discipline, or centralized, regimented, and standardized structure, a state-of-the-art computer system won't give a company comparable success.

The information contained in integrated computer systems is useful not just in the core operating processes. Operationally excellent companies are passionate about measuring and monitoring to ensure rigorous quality and cost control. They generate detailed data with which to make management decisions.

Operationally excellent companies have aggressively pursued mobile technologies to extend their control and to improve customer service. The hand-held computers used by Hertz, FedEx, and UPS employees are examples. Companies like L.L. Bean and Lands' End have driven teleservices technology to new levels of sophistication. What these companies find at the leading edge of technology is better operational efficiency and control.

CUSTOMER SERVICE

Today's operationally excellent companies have revolutionized their business in another dimension that Henry Ford never imagined—customer service. Poor service can add substantially to a customer's total cost through wasted time and frequent errors. Operationally excellent companies return this lost time to customers.

Most companies that buy from other companies maintain large accounts payable staffs not because they can't redesign the process to automate payables, but in essence to insure themselves against the inaccuracy of their suppliers' invoices. Consumers, too, maintain the habit of always spending some portion of their busy lives correcting others' mistakes. The rule for operationally excellent companies is: If you truly want to have the lowest total cost, make sure your service is effortless, flawless, and instantaneous.

To do this, operationally excellent companies have redesigned the customer-service cycle, aggressively streamlining the selecting, ordering, receiving, paying for, and maintaining of a product. Just think of how much easier the car rental process is today than even a few years ago. Hertz #1 Club Gold members record their car rental preferences only once. When they place an order, either through Hertz or with a travel agent, all the information Hertz needs is already entered. Club members encounter no lines or delays at the airport. They get right on the bus and are dropped at their cars. Hertz automatically bills their accounts at the end of the rental. No fuss, no hassle. That's the standard of basic service in an operationally excellent company.

One of the main keys to achieving operational excellence in customer service is the same as in manufacturing: Do it one and only one way. Here again, variety kills efficiency.

Another key is getting customers to adapt to the operationally excellent company's way of doing business. McDonald's provides the classic example. Your mother couldn't get you to clear your dishes, but McDonald's did. When underneath the Golden Arches, you expect to follow some implicit rules: Bus your own table. Stand in line to give your order. Know what you want when you get to the front of the line. Don't ask cashiers to hold the pickle—take it off yourself. McDonald's has built a whole system of social norms that patrons readily enforce among themselves. Of course, customers comply only because they get low prices and efficiency in return.

Likewise, Southwest Airlines makes a selling point of not providing food, advance check-in, and baggage handling. It explains that such frills preclude low prices and impede reliable service. Customers have bought the message and adapted their behavior accordingly. Going on a trip? Pack light, says Southwest, because you'll be carrying it. Hungry? Grab a bite before the flight. Just arrived? Step into line for a boarding pass.

Why are operationally excellent companies uniquely qualified to deliver superior basic service? Because they enjoy three advantages. The first is focus, making hassle-free basic service a key part of their unmatched value proposition. Second, their operating models support efficient, zero-defect service. The practices of operationally excellent companies are part of the rule book for zero-defect service. Third, they have effectively exploited information technology to redesign basic service tasks. Information technology has made service available anywhere at any time. It is the catalyst of the service revolution.

EXPLOITING THE VALUE
LEADERSHIP ADVANTAGE

As operationally excellent companies create an unmatched value proposition of best total cost, the question comes up: What's in it for them? There is only one answer—growth. Other market leaders might raise prices to exploit their product advantage, but such a tactic runs counter to the operational excellence strategy.

Wal-Mart could raise its prices a measly 1 percent tomorrow in an effort to add $800 million to the bottom line. But it won't. If it raised prices to exploit its current advantage, it would merely be stealing from its future success. Higher prices mean less value leadership. Less leadership means lower growth and, ultimately, shrinking margins.

Operationally excellent companies obtain their growth in three coordinated ways. They work to assure a constant, steady volume of business so as to keep their assets continually working; they find new ways to use their existing assets; and they replicate their formula in other markets. Let's take these growth generators one by one.

Having invested heavily in stores, plants, airplanes, or other fixed assets, operationally excellent companies know that using those assets as many hours as possible every day boosts both revenues and financial returns. They also know they must smooth out demand fluctuations as much as possible to avoid inefficiencies that come from a boom-and-bust cycle of volume. Consequently, they strive for large, consistent volume throughout the day, the week, and the year. Demand peaks and valleys become operational problems to be "managed," because nothing undermines efficiency—and hurts low unit costs—faster than slack in a system. Machines sputter to a stop; workers cool their heels. In short, fixed costs keep adding up, while no product flows to pay for them.

Two pitfalls, however, plague demand management aimed at keeping capacity utilization both high and steady. The airline industry illustrates one. Yield management has introduced so much complexity into that industry that it may be increasing costs, not lowering them. The airlines hurt themselves by using different kinds of aircraft for different kinds of routes, and by offering different fares to attract different kinds of customers. Remember, variety, even variety in price, destroys efficiency. The second pitfall is that using price and promotion to shift demand can teach customers to withhold buying pending a special deal. When that happens, demand actually becomes lumpier. Wal-Mart sidesteps this pitfall by rejecting promotional spending in favor of everyday low prices. Promos, ads, and sales are the brainchildren of product-marketing people who don't understand the monkey wrench they can throw into retail operations. No wonder companies like Kellogg, General Mills, and General Foods rank far down the list for operational excellence. Their behavior saps efficiency.

A second way that operationally excellent companies fuel growth is by finding different markets to penetrate with their existing assets.

Recall how McDonald's created a breakfast market. Aware that much higher profitability would flow from filling its restaurants at more than the lunch and dinner hours, it launched a campaign to convert the hamburger crowd to born-again breakfast eaters. The idea was to get customers to drop by first thing in the morning for a cup of coffee and an Egg McMuffin. With the success of the campaign, McDonald's boosted the utilization of its fixed assets—its restaurants and equipment—to 10 to 12 hours a day from just six or eight. This doubling slashed unit costs and overhead and gave a big lift to the bottom line.

FedEx took another approach to growing demand while shrinking the idle time of its assets. Its planes once flew almost entirely at night and sat unused most of the day. FedEx contacted L.L. Bean, which at the time was shipping its clothing and camping gear by any available carrier. FedEx offered to take over all of L.L. Bean's shipping. Urgent deliveries would still take to the air on the night schedule, but the rest would fly by day—using the idle planes. This arrangement, FedEx argued to Bean, would both take up its own slack and give the retailer faster service. Bean agreed and a partnership was born.

The other route to growth for operationally excellent companies is through replication—transporting an efficient, standardized service to a new location. Wal-Mart has one formula, FedEx another, Southwest Airlines yet another—and the buying public firmly associates those formulas with the companies' brand names. So once that formula is defined, perfected, and widely known, an enterprise can fire up its cloning machine. When it opens a new outlet, a company doesn't have to slog up a steep learning curve. It simply replicates the processes it already performs. Nothing repeats like a success.

McDonald's expertise in opening new restaurants around the world—something on the order of more than 1,000 a year—contributes substantially to its remarkable success. In countries like China and Russia, where most people once thought a hamburger was someone from a certain German city, millions now know and love the familiar Big Mac formula. The more than 9,000 U.S. restaurants are now complemented by almost 5,000 Golden Arches around the globe.

Formula replication can work well not just in the service sector, but for product businesses as well. Highly successful Nucor Steel, for example, replicates its mini-mills in various parts of the country, following the same layout and procedures. Toyota and Honda have transferred their efficient methods and even their management style—using virtually identical designs and procedures—from Japan to Kentucky. Although the workers are American, the plants run on the once foreign-seeming Japanese model, and nothing seems lost in the translation.

FORMULA! FORMULA! FORMULA!

What then distinguishes operational excellence from operational competence? Hard choices: less product variety; having the courage to not please every customer; forging the whole company, not just manufacturing and distribution, into a single focused instrument. The operationally competent company will shy away from these tough calls—and pay the price.

A canny weave of unparalleled know-how, technology application, and tight management—that's what makes a leader in operational excellence. The secret of succeeding with this value discipline is summed up in a single word: formula. Formula often has a negative connotation, but for operationally excellent companies, it's the foundation for an aggressive and highly successful enterprise.

5

ONE COMPANY'S EXPERIENCE–AT&T'S UNIVERSAL CARD

CHAPTER 5

ONE COMPANY'S EXPERIENCE–AT&T'S UNIVERSAL CARD

So intent on efficiency was Paul Kahn, the first president of AT&T Universal Card Services, that by the day his operation began in March 1990, AT&T computers had already screened most of the U.S. population for credit worthiness. That meant that AT&T phone reps could instantly approve cards for any caller who could be matched in the database. "We didn't need any other information," he says.

On that first day, AT&T reps, based in Jacksonville, Florida, rode out a tsunami of calls. And the database matched the vast majority of them. "The card was issued the next day," for credit worthy customers, says Kahn. "It was out the door and in their hands in a week." Competitors often took a month or longer to issue a card.

With such operational excellence, Kahn would permanently change the rules of competition in the credit-card business. His idea—simple to describe but hard to execute—was to run a lower-cost operation that could deliver gilt-edged service unheard of in the credit-card industry.

The story of Universal Card, which we tell below with the help of a roundtable of comments from Kahn and several current employees, shows in detail the power of the operational excellence value discipline. By building a new business on that discipline, Universal Card leaped in a matter of months to market leadership. So efficient was its new operating model that it prompted fierce competition in the industry. The competition in turn squeezed profits and spurred consolidation. The top 10 card issuers in 1986 had a 33-percent market share (by card volume); by 1993, they had increased their share to 40 percent.

What prompted AT&T to think it could rush in and establish a winning business? Partly the juicy profit margins. Retail banks then ran most credit-card operations, and historically, credit cards had been the banks' most profitable product, returning up to 5 percent on assets as compared to the 1 percent typically earned by a well-run loan portfolio. Kahn was bent on taking a big bite out of those profits.

Also drawing AT&T into the fray were changes in technology. New non-bank competitors could install inexpensive, off-the-shelf computer programs. In essence, they could operate like the Wal-Marts of the credit-card industry, undercutting traditional providers with a radically lower cost structure.

AT&T also bet it could build business by leveraging brand awareness, which hardly existed in the credit-card industry. Half of credit-card customers didn't know that their cards were issued by one of 6,000 organizations. To them, it was just a MasterCard or Visa. AT&T had a venerable name to offer.

Perhaps most important, AT&T could see that customers weren't satisfied with the cards they had. Card users didn't like the high annual fees, and with one or two exceptions, they didn't like the mediocre service they received. Many people were downright angry about surly phone representatives who took weeks to process a simple change of address, or stolen-card bureaus that couldn't issue new cards without embroiling customers in a tangle of paperwork and phone calls. AT&T figured it could deliver much better service.

The world was changing in 1989 in a way that would hurt traditional issuers with or without the coming of new competitors. Their customer base was finite, and their product was mature. All of a sudden, card issuers couldn't command high fees while delivering lousy service. Their customers were ripe for a better offer, and AT&T made it.

AT&T bet right. The AT&T Universal Card Services operation was profitable in 27 months, way ahead of schedule. Today, it's the second-largest banking card in the industry. Over 20 million domestic card holders, representing receivables of over $11.1 billion, use it as a combination long-distance calling card and general-purpose credit card. They get the same benefits they would with other premium credit cards but with no annual fee, low interest charges, and a discount on long-distance telephone calls.

Recounting the UCS story, which began just over four years ago with a concept team of 35 people, is Paul Kahn, its first president, who has since become president and CEO of Safeguard Services. Joining Kahn are four current employees: Jim Kutsch, vice president of Computing and Networks; Linda Plummer, a senior manager of Customer Relations; Daniel Patterson, a member of the bilingual team of telephone associates; and Bridgette Waters, a member of the Consumer Protection Department.

THE PROPOSITION OF OPERATIONAL EXCELLENCE

When AT&T brought the Universal Card to market, it had laid plans to succeed twice over. For starters, it priced the card lower than any other available—it was free, and the low interest rate could save heavy card users hundreds of dollars a year. Customers jumped to get such an appealing card to slip in their wallets. AT&T followed up with no-hassle, responsive service that got customers to pull the card out of their wallets more and more often—and competing cards less and less. By getting people to acquire and then use the card regularly, AT&T met its key objective of capturing a high share of the outstanding credit-card debt. AT&T had loosed a world-beating formula for ravaging the business of almost all its rivals—and it took competitors years to figure that out.

> Paul Kahn: Let me try to bring the Universal Card story into focus. When we launched the card, we launched it with very clear, simple marketing messages. No fee. Free for life. Our statement was very powerful; we combined a calling card and a credit card that lets customers get rid of the two cards in their wallet and substitute one. We simplified life. But we also added value; every time customers use the card for long distance they get a discount.
>
> I think the key to success for AT&T Universal Card was running an operation according to the basics of quality, service, value, and treating people decently in an environment that allowed them to be successful—particularly while our competitors didn't have that environment. That gave us our

competitive advantage. Unless you have a new product, unless you have a new positioning, you've got to worry about how to get competitive advantage. We did that by running an excellent operation that delighted the customer with superb service.

As time went on, because we had variable interest rates, our rates kept dropping with the prime rate. Our rate dropped five times over two and a half years, while our competitors' fixed rates stayed the same. Every time the prime dropped, we would tell consumers. We would tell prospective customers to transfer their balance from our competitors to us to save money. We kept hammering home a simple message: value, quality, service, and a company you could trust.

We provided excellent service by designing everything that was touched by a customer to be better. That meant gearing our people to handle customer problems by phone, not to ask customers to fill out some lengthy form. It meant eliminating errors in our paperwork processes. It meant streamlining the process for delivering cards. It even meant mundane things like redesigning our statements. We designed the statement and then redesigned it again to make it easier to read and easier to send in payments. Whenever we found a problem, we fixed it. We maintained a continuous focus on fixing anything that went wrong.

Achieving perfect quality, no errors, is not only doable, but it's a valuable competitive strategy. For example, every month we took in eight million payments. If we had a 1 percent error rate on those eight million, that would mean that 80,000 customers a month would be upset. They would be upset that their payment didn't get posted to their card, or that they got a late fee, or that interest was charged. That's an unacceptable number. Eighty thousand times 12 equals a million customers a year who would get upset over one of our processes. We started at about a 99.8 percent error-free payment-processing rate and got to 99.9 percent.

Our service platform was also designed to be faster than any of our competitors'. For example, we wanted a turnaround time of three days from the time customers called us to the time we had a credit card in the mail, and seven days for a card to be in our customers' hands, with the average postal delivery time. We met that goal, except for a couple of start-up months when the volume was overwhelming.

Our intent was to design a system to delight the customer—not satisfy, but delight. If customers had a problem with a card, or a change of address, we changed the address online instead of sending out two more bills with old addresses on them. Most competitive companies in the United States today are geared towards customer satisfaction. They haven't looked at the tremendous competitive edge you get by delighting customers.

COMPETITORS: SLOW IN THEIR RESPONSE

History now shows that AT&T Universal Card Services' formula was a winner from the start. Customers were the first to realize it; the competition took longer. For months after the card's launch, other card issuers heard the same marketing messages from UCS as did everyone else, and they heard the same reports of stellar Universal Card service. Still, they remained dubious and hesitated to react. What they couldn't seem to hear was the sound of UCS busily sawing the legs off the structure that supported their obsolete business practices.

What finally got competitors' attention was the sucking sound of market share rushing south to UCS's Jacksonville, Florida, headquarters. Although most card issuers would get returns of 15 to 20 percent on their direct-mail campaigns, UCS got a phenomenal 40 percent, according to Kahn. As the months clicked by the evidence became overwhelming that UCS was gobbling competitors' shares.

Paul Kahn: We had American consumers moving from our competitors to us. Our competitors waited for two years to react. So for two years, we took market share. Our competitors never got it back. They thought we couldn't sustain our success. They thought we were a flash in the pan. But we were not just taking share; we were taking their best customers, and they realized this trend was going to break them.

They kept getting higher-risk customers, and we kept taking the lower-risk customers. They didn't understand that we had done a paradigm shift.

AT THE CORE OF THE
BUSINESS IS OPERATIONS

Operationally excellent companies like AT&T Universal Card Services depend for their success on processes honed to deliver high levels of reliability at low cost. The processes may include issuing cards, changing addresses, or correcting billing mistakes. Standard operating procedures, practiced over and over and tested rigorously to isolate glitches, keep the costs of each of these procedures down. Note that companies like AT&T Universal Card Services don't ask their people to make up new rules as they go along to deliver great service. They ask them to follow established procedures that, over and over again, have yielded the hassle-free, unexpectedly good service that customers value.

> Paul Kahn: Our focus in lowering costs was to cut out waste and errors. According to statistics, 15 to 20 percent of any industry's costs come from error, recovery from error, and from cleaning up mistakes. So we decided that waste was where we wanted to cut costs.
>
> We were not going to reengineer in the way some companies are doing it, which is just to lay off a bunch of people. We were going to drive out errors and drive out defects. That's how we were going to lower the costs. Then we'd either pass the savings on to the consumer or put it into our bottom line. And frankly, we did both. That is why we could offer the card with no annual fee, drop our interest rate, and still come up with profit margins that were equal to those of other people in the industry.
>
> We didn't want to lose customers, either. As everyone knows, it's a lot more expensive to get new customers than to keep the ones you have. In our business, it takes a year and a half to get customers profitable. At that point, the longer you keep them, the more profit you make. In fact, it's about $35 a customer in annual profit. So we wanted to keep our volun-

tary attrition rate to a minimum. And in fact our attrition rate at UCS was less than 2 percent a year. Citibank's was 15 percent. So they lost a customer that they basically had a six- or seven-year life with. It's too soon to tell, but AT&T certainly hopes for a much longer tenure.

Our attrition rate was partly determined by how we handled customer complaints. Because we took an approach of delighting customers, we had about 200 customer complaints a month coming in, whereas we had 2,000 customer compliments. That's pretty good performance, because American consumers do not like to write compliment letters. They typically write only when they're irritated.

Daniel Patterson: I try to give the same level of service to everyone; there is not any one person to whom I give better service. The extraordinary is really the ordinary around here. I was recognized with an award after I took care of a case that involved a couple that had flown to Mexico. The wife fell off a horse, broke her hips, and required surgery. The doctors in Mexico would not accept a credit card, nor does medical insurance here transfer over into Mexico. So the couple was in a bind. With the help of a co-worker, we were able to reach the doctors and tell them we would deliver cash on the spot if necessary. After we faxed some information, they accepted the card. Following the surgery, everything was fine. We helped the couple get reservations for their return home. That was a great feeling. The heroic is a part of everyday work.

Kahn: Let me give you another example. One Saturday in September, back-to-school time, we had a computer failure. The failure was at one of our vendors. As a result, about 40,000 of our customers were turned down when they went to use their credit card to make a purchase. Now that's not a very nice experience, particularly if you're in a restaurant trying to impress some friends and the proprietor says, "Sorry, this card is no good." What was worse was that some overzealous merchants, about 200 of them, actually took the card away from the customer and said, "I'm sorry, we have to take this card away from you." Merchants get a $50 reward if they retrieve a fraudulent card, and a few decided they had an opportunity to make some money.

So here we have a lot of dissatisfied customers. We didn't know anything about it until customers started calling at about noon on Saturday. But we had developed an infrastructure in our company for that kind of thing. With our online systems, we could find out in short order that there was a trend going on. What did we do about it? We got to the vendor and got him working on the problem, which took about four hours to fix. We also put out an online message to our reps instructing them, when any customer called, to (1) apologize for the problem and (2) find out where the problem took place. If the customer was at the merchant, the reps were to get the merchant on the phone and authorize the transaction. If the customer was not at the merchant, then our reps were to get the merchant's name. They were to tell the customer that we would call the merchant, apologize on behalf of them and ourselves, and tell them what the error was.

On the following Tuesday, we had management swing into action. I sent out letters with my signature to 40,000 people. I explained what happened, said that it was a computer error, that we were very sorry we had inconvenienced them, that we were going to make sure it didn't happen again, and that, here, please accept a $10 gift certificate to use in any way you want. The next month we got more compliments than we had ever received before. People were not used to being treated that way. But we had positioned ourselves differently, and the way we reacted showed that we meant what we said—it really solidified our relationship with our customers.

MAKING THE CORE PROCESSES SHINE

Companies that derive their success from operational excellence live or die by process improvement, governed generally by the principles of total quality management. AT&T Universal Card Services is no different. It constantly polishes its standard operating procedures to cut costs for itself and cut hassle for its customers.

Paul Kahn: We developed a process called continuous improvement. In fact, it became a core value of our company—whatever we're doing today we can do better tomorrow. Inside the company, that approach diffused negative emotional reactions. Typically, you criticize somebody for making a mistake. Instead, we took the position of saying, "You've done something wrong in the spirit of continuous improvement—how can we do this better the next time?" We always tried to be forward thinking, asking, "How can we get better?"

We made the quality improvement process a little bit of fun. We actually stole an idea from the FBI, and we created a "10 Most Wanted List" of quality defects. Every department at UCS had a "10 Most Wanted List" that it put up on a wall. Teams were assigned to work on each item on the list, and when they corrected one of those defects, we would have town meetings with every employee, and we would bring the teams up and applaud them and give all of them ceremonial plaques. At any given time, we had about 125 major quality improvement projects going on, all of which had teams assigned to them. When one improvement project was retired, the next one went up on the list. It was a continuous process.

If you ever get complacent and think that you've achieved high quality, then you really are at the end of your quality journey. One of the things I did was create an environment that forced everyone in the company to be closer in touch with our customers. Every manager in the company is required to listen in on customer service calls two hours per month. None of us were competent at handling them, but we had to at least listen and understand what people wanted.

In June 1990, three months after we launched, we thought we were pretty hot, I have to tell you. We had just worked out our telephone problems, and we were pumping cards out like crazy. But we benchmarked ourselves against the Baldrige template, and we found that we scored only 150 points on a 1000-point Baldrige total. It was a rather humbling experience to me.

When we won the Baldrige Award in 1992, we scored somewhere between 700 and 750 points, which is where you win the Baldrige. So in three years, we had moved up significantly.

EMPOWERED BY INFORMATION TECHNOLOGY

Enabling AT&T Universal Card Services' success is information technology. No surprise. Real-time, hassle-free service in today's world comes only through the speed and integration of computers and databases. Not only is AT&T Universal Card Services highly automated, however. Computer systems essentially define people's work—they proceduralize it. In effect, the system manages the process.

> Paul Kahn: We used technology very much as a competitive weapon. What we wanted was a "high-tech, high-touch" environment. We started with a database of all customers in the United States, so that if you called up, and we had your phone number—bang!—we could map you against the database, have you profiled in front of the telemarketing rep before he or she ever answered the phone.
>
> Our phone response rates were immediate, as opposed to two, three, or four seconds. By cutting that wasted time, we could afford to give more service to our customers.

> Jim Kutsch: We draw an analogy between UCS's computer screens and those of a modern jet fighter where many of the lesser functions are automated so the pilot can focus on the main job. We want to do the same thing, taking care of the little things behind the scenes, providing the right tools and facilities, so that our telephone associate can bring full attention to his or her conversation with the customer.

> Kahn: Every transaction we looked at, every time we touched a customer, we tried to figure out how could we do it better and faster. For example, we used expert systems so that ordinary people with no credit experience could grant credit-line increases. We made it simpler. UCS is now on its

fourth iteration of this workstation system, so that gives
you an obsolescence life of hardware and software of about
one year, to give you an idea of how we tried to keep a competitive edge.

Kutsch: What we have evolved is the U-WIN workstation—our customer service delivery platform. Our PBXs
(local phone networks), our call-management system, our
800 services, our network—all of these have also been advan-
tageous. We think our database analysis is particularly powerful. We've collected data since the launch of our product and
have details of every transaction. This gives us a tremendous
asset for reviewing and improving our operations.

THE PEOPLE EDGE

The other tremendous asset at AT&T Universal Card Services is a
motivated work force. The company has managed to accomplish what
few other companies seem capable of: It has created a high-spirited attitude in people performing routinized work. Its approach has been to
hire eager people with a lot of potential, train them to view excellent
service as routine, and assemble them in teams to solve problems that
lead to continuous improvement.

The efficiency of UCS, like other operationally excellent companies,
also stems from a high degree of role specialization combined with
broad integration. Consequently, associates, such as those on bilingual
telephone teams, can handle, or find someone nearby to handle, most
problems.

UCS' formula for managing people builds a degree of organizational
strength that most competitors find unassailable. Training helps people
perform; performance boosts self worth; a sense of worth builds
employees' loyalty to the team and customer. The benefits of this self-
reinforcing approach to management show up in both happy employees
and a robust set of profit figures.

Linda Plummer: One of the philosophies we have at UCS
is that if we have satisfied and delighted associates working
for the company, then we will have satisfied and delighted

customers. We are all on the same team. Our motivation is to be the best employer in town—to provide our associates with a working environment that they enjoy.

Jim Kutsch: We have a very empowered work force. The expectations are clear; our objectives are not ambiguous or up for interpretation; and we aren't tremendously title conscious. There is a great deal of pride and enthusiasm here, and that is also a motivational factor for individuals.

Bridgette Waters: The company makes you feel you can talk to them—the open door policy really is in effect. You don't call our president Mr. Hunt, you call him David. He stopped by my desk one day, asking me how I was doing. He's very down to earth; it makes a difference. You don't get the feeling around here that someone is better than you or you are beneath them.

Kutsch: We work together to improve, and we have a lot of cross-functional teams to get things accomplished. In some cases, a cross-functional team may be set up to address an issue of quality improvement, based on the 120 daily indicators of quality that we measure. The teams come into existence based on a particular need that we perceive in the company. I think the cross-functional approach builds a sense of solidarity in the company.

Daniel Patterson: In addition to cross-functional teams, UCS organizes its associates into teams within a specialized function, such as the bilingual team. My team ranges from 15 to 40 associates; other specialized teams can be smaller. Our environment is such that you can hear what other team members are saying. If someone is in trouble, with a screaming customer or a tricky procedure, we all pitch in and try to help.

Plummer: We have quality people in part because we have an extensive employment process here. It's expensive, but we have found the quality of people we get is just out-standing. They are very motivated, bright, and after they've gone through the entire process, they feel as if they've really made it.

We begin with a written, problem-solving test, which is not designed for any specific level of education—you're not excluded if you haven't been to college. Those who pass go on to a role-playing situation: We give out information about a fictitious company, then someone comes on the telephone and pretends to be a customer. We listen to this interaction and evaluate people in terms of flexibility, interpersonal skills, and general approach to the customer. Qualifiers go on to a formal employment interview, where we look for such things as stability of past employment, satisfactory work record, and so on. Another interview follows, with an operations manager. We take candidates through our operations center to give them a feel for the environment. We also conduct a drug-screening test and a reference check. So it is a pretty thorough process; out of 100 people who take the written test, we end up hiring about 10. A few years ago, I was the employment manager here, and I think we tested 13,000 people that year.

Patterson: For me, getting hired was a long process. I first applied three-and-a-half years ago when UCS was doing massive hiring. The company was renting out a convention center and testing 200 to 300 people at once. I took the test, and then took a phone test to see how well I could communicate with cardholders. Then I had an interview. It took a while to get hired, but it was a good experience. There is a lot of competition—one thing that impresses me is that everybody here is very well educated. The competition, when another job is posted here, is very tough. It makes you a lot stronger and better.

Plummer: Once people are hired, they go through a training class that lasts five to six weeks; then they get another two weeks of on-the-job training. We have certified instructors in the classroom, and coaches—one coach for about every three people. Students listen to associates taking calls, and take calls themselves with an associate monitoring them.

One of the things I like about the training is a module called "heroes"—examples of UCS people who have won awards for going above and beyond in helping our customers. This starts people thinking about how we try to give extraordinary service, and how our associates are empowered to make decisions in challenging situations. We're proud of the fact that our training program has been evaluated by the local community college here, which gives three credits to students who complete the course.

Patterson: Once you're put on a team, there is almost continuous training in new methods and procedures and system updates. If you're not on the phones taking calls, you are involved in other projects or in training. The training is extensive.

Waters: If you are successful at what you do, your ideas get used to help someone else. If someone's stats are not as good as mine, I'll go over and ask to see what they are doing, or they might come and observe how I handle a situation. It could be a simple adjustment that could make a difference in their performance, and even affect their annual review. It feels good—you can help someone, and someone can help you.

MEASURE AND REWARD

Companies that climb to the pinnacle of operational excellence get there by sweating the details. They have to assure that not just their strategy, but its execution—a matter of endless small tasks—verges on the perfect.

The only way to wrestle every small, necessary task into its proper place is to track it, measure it, and reward people accordingly. Universal Card Services compiles hundreds of statistics to keep tabs on its performance, and every employee gets incentive pay for keeping the statistics in top shape. Added to the incentive of pay is a panoply of programs that assure that no good deed goes unrecognized.

Paul Kahn: Now when all is said and done, in order to deliver top quality, you have to measure it. You have to measure everything in your organization. And we did exactly that. We had 350 different processes, and every one of them was measured. About 120 of them were considered critical measures, and, in fact, we tied everyone's compensation in the company to achieving those quality objectives every day. Not monthly, not weekly, every day. We had to meet 96 percent of those quality objectives, and if we did, everyone in the company, from myself all the way down to the janitor, got a quality bonus in their paycheck.

Jim Kutsch: These 120 indicators tell us how we're doing at producing world-class, pre-eminent customer service, day after day. The indicators range from the length of time it takes us to answer the phone to whether a billing statement was mailed in a timely fashion. We also have indicators of internal systems: Was our computer network available? Was the response time adequate? Were any telecommunications components down? In our information technology group alone there are nearly 50 such daily indicators measured, the majority of which deal with day-to-day production, support, the nuts and bolts of running the system. Other indicators include courtesy and professionalism on the phone, timeliness in responding to a customer request, and so on.

Kahn: We started measuring daily quality at UCS from the first day of business. Specific indicators varied over time, in response to changes in processes throughout the company. We also continuously raised the threshold against which we passed or failed a particular indicator—so it's harder to achieve an indicator today than it was three years ago.

By listening to customers, we actually got to a point where we knew the eight things that customers want in a credit card. We had each of those weighted, so we knew by weight what was important: price, service, ease at point of sale, company trust, and so forth.

Every month we called up our customers and asked them how we rated in these catagories. We also asked how our key competitors rated, such as American Express, Discover, Citibank, and some of the others. So at any given time, we knew the most important things that our customers wanted, and how we were doing and how our competitors were doing. It was nice that typically we were ahead in six, seven, occasionally in all eight categories, but more important, when we saw a lag between ourselves and our competitors, we could go back and say, "What is it they're doing that we're not doing, and what can we do to improve?"

We eventually got even more sophisticated. We actually could model how much increased volume we would get if we increased by 10 basis points a certain category of customer satisfaction. It got us to focus and concentrate on the key things that our customers wanted, and gave us the ability to differentiate ourselves from our competitors.

Linda Plummer: There are many ways of recognizing employees at UCS—about 40 programs in all. We have "Associate of the Month," which rewards people either for their work in the company or the community; we have "Service Excellence" awards, which reward those who provide outstanding customer service; and we have the "President's Circle," given to those who best exemplify the values our company is based upon. About 3 percent of our associates receive the President's recognition, which is determined by the president or his team members.

The "Power of One" award is one of our most significant. You can nominate anybody who goes way above the call of duty. The nominations are anonymous, and go to the chief operating officer, who shares them with his vice presidents. They vote, and then go around to the associates' desks with an air horn to announce the award and present it with a check. It is really nice.

Kutsch: In our world, heroic efforts are an everyday occurrence. It is not just the stray, occasional story. We have customers who have lost their card in an ATM machine, or had their wallet stolen and are stranded, where heroic efforts have been made by our associates. But we don't view them as such—this is important. We want world-class, pre-eminent customer service; we want to exceed expectations; and, as we are constantly emphasizing around here, we want to delight the customer. And we want it to be routine.

THE VALUE OF DISCIPLINE IN OPERATIONS

Today, AT&T pulls in an average of 300,000 new accounts per month, and that growth in card holders stems not just from the original sales pitch. It comes from a finely tuned operation—from technology that speeds performance and arms phone reps with leading-edge tools; from continuous-improvement practices that squeeze cost and waste out of the company's processes; from a broad set of measures that lets managers monitor performance and compare it with the performance of competitors; from a comprehensive set of programs that assure that associates and managers know what customers want; and from a workforce hired, trained, and rewarded for self-starting, heroic performance.

As further evidence that it's doing something right, AT&T Universal Card Services, employing 2,800 people who field more than a million calls each month, in 1992 won the Malcolm Baldrige National Quality Award for outstanding performance in customer service. This in UCS's third year of operation!

6

THE DISCIPLINE OF PRODUCT LEADERS

CHAPTER 6

THE DISCIPLINE OF PRODUCT LEADERS

With an amazing 1,300 inventions and 1,100 patents to his credit, Thomas Alva Edison epitomized the lonely inventor. What few people realize, however, is that he bestowed on mankind a gift far greater than any single invention: the process that underpins product leadership.

In 1879, Edison pioneered what was probably the industrial world's first effective product development process, a means of generating invention after invention and applying those inventions in useful, commercial products. His Menlo Park, New Jersey, lab, where he and his bright assistants collaborated, became the model for the labs of today's product leaders such as Microsoft and Sony.

Edison, his product development process picked up and polished like a jewel by so many companies, would surely be thrilled at the prodigious output of product leaders today. He would cheer their obsession to create products that incite a passionate customer response. Indefatigably curious and driven himself, he would celebrate employees' fixation on breaking new ground, on cracking the code of confounding concepts. And he would admire their nose-to-the-grindstone creative determination and their persistence in turning inventions into products. After all, it was Edison who said, "Genius is 1 percent inspiration and 99 percent perspiration." It was also he who said that he knew 50,000 things that wouldn't work—after he had sweated through innumerable failed experiments to develop a device to store electricity.

But at the same time, Edison would be appalled at the sad state of product development at many companies today. Too often, firms forsake invention for its poorer cousins—refinement, repackaging, and reformulation.

Many products exiting R&D labs today come tagged as "new" and "improved," but they don't deserve either label. They're not new breeds; they're mongrels. And while they may be different in some trivial way, they're hardly improved. Instead of exciting customers, they bore them.

How about these for yawners: Nabisco expands its Oreo cookie line by stocking store shelves with mini Oreos, Double Stuff Oreos, larger packs of Oreos, smaller packs of Oreos, and seasonal packs of Oreos. The result: Sugar-hungry shoppers' eyes glaze over while Nabisco sees negligible growth. Ford and GM bring out one car model after another that all look the same. To replace the Escort, Ford creates...the Escort, with barely discernable improvements. How do car buyers react? They look to Europe for leadership in styling. PepsiCo tiptoes to market with "innovations" such as freshness date stamps on soda cans, and Coca-Cola tries to strum customers' nostalgia strings by bringing back old corset-shaped bottles—this time in plastic. Cola buyers reach for Snapple.

It's no wonder consumers easily sleep through such product innovation. What have these companies offered? Features that don't inspire. Variety without benefit. Frills without meaningful value. Incremental changes that are easily overlooked. Customers are too savvy to fall for this desultory stab at razzle-dazzle.

What wakes customers up—and would have delighted Mr. Edison—is product leadership, a company displaying the ability and determination to make products that customers recognize as superior—products that deliver real benefit and performance improvements. This is a hard lesson to learn, and many companies learn it only after suffering through customer indifference to their boring products. Nabisco, for one, has come back like a powerhouse—its SnackWell cookies are the best selling in the nation. Ford has won big with its breakthrough Taurus line and hopes to score again with the new Windstar minivan and its global car, the Mondeo.

Ford, Nabisco, and others got the message: Customers aren't impressed by one-time innovations followed by countless "improvements." To be product leaders, companies have to show that they can create a steady stream of standout products that will keep customers awake with anticipation—products that turn people's heads and make their hearts beat faster.

Microsoft introduced Microsoft Office, for example, a bundle of software that enables customers to automatically update figures in three applications—Microsoft's Excel spreadsheet, Word word-processing and Powerpoint presentation packages—while working in only one.

Thermo Fibertek Inc. brought to market a new process for extracting printer's ink from used paper, which enables paper producers to make clean, white, recycled paper stock never before available.

Johnson & Johnson's endoscopy division has roared to market leadership in just a few years. Born of Ethicon's suture division, they have pioneered less invasive surgical techniques that, among other things, cut the need for sutures!

A hundred years ago, of course, Mr. Edison had narrower goals than inventors of today. He concentrated on the utilitarian benefits of his products. His practical bent was similar to that of Rubbermaid, which recently introduced a mop bucket formed from plastic that resists microbial growth, and a clever new mailbox equipped with a little flag that pops up automatically after the mail comes and saves homeowners from making futile trips.

Today's product leaders find that customers have a much broader perception of performance. They crave a mix of tangible and experiential benefits. They expect the performance of breakthrough products to move their rational and their emotional selves. For instance, Harley-Davidson has already received unsolicited orders for its 100th anniversary edition motorcycle even though the bike won't go on sale until 2003. There are 200,000 members in the Harley Owners Group, which sponsors bonding rallies where members compete in H.O.G. games. Harley customers buy more than a motorcycle; they buy a lifestyle.

Indeed, for some products, experiential or emotional impact is a prime measure of performance:

■ Nike, Reebok, and Swatch products indulge people's hunger for an association with sports heroes, the rich and famous, or peer recognition—in the same way that Revlon sells hope, not cosmetics.

■ Maxis computer games make people masters of the universe, letting them pull the levers of power to run a metropolis, command a military base, publish a major newspaper, or, by turning them into an insect queen, direct an ant colony.

■ MTV, Walt Disney, and directors like Steven Spielberg create products that go beyond supplying fresh concepts and stimulating experiences; they actually change our culture.

THE OPERATING MODEL
OF PRODUCT LEADERSHIP

No matter what their formula for delivering features that affect performance and experience, product leadership companies deploy an operating model based on many of the same principles developed by Thomas Edison.

Edison began by creating an inspiring vision of each new product before development work ever started. He believed he needed to fuel his organization with the dream of improbable achievements. Like product leaders today, Edison saw breakthrough products as the sustenance his enterprise needed to keep on working. Generating a stream of products that are clearly better and possibly pathbreaking is what injects passion into product leaders' organizations and keeps them alert and alive.

Edison maintained a very fluid organization. He regularly reorganized and redeployed his human resources toward the most promising projects.

His business wasn't driven by procedure, but by the extraordinary talents of key individuals who developed and marketed breakthrough after breakthrough. Every step along the way was focused on realizing the vision. Edison structured his employees' jobs around the creation of products, not around any particular function.

Edison knew he didn't stand a chance of reaching his goals without hugely talented people, competent in their disciplines, naturally curious, and energized to tackle nearly impossible objectives. He came close to cloning himself at Menlo Park, developing disciples who could think and act in his inventive, efficient manner. Product leaders today also realize that people of superior talent who are capable of ambitious, right-to-left thinking are crucial. It's not by accident that well-known product leaders such as Microsoft, Disney, Fidelity, and Glaxo are among the most aggressive recruiters on top college campuses.

Along with breakthrough visions, Edison, like product leaders today, harnessed the motivating power of ambitious targets. Edison wasn't the first one to create a light-emitting device; several of his contemporaries had engineered contraptions that would shine for a few moments. But Edison's target wasn't a light bulb; it was the ability to light a building or even a town. That lofty target channeled his creativity. He then worked backwards from the goal to figure out the steps required to achieve it. Working right-to-left, as we call it, is a pervasive characteristic of successful product leaders. They shun aimless experimentation and daydreams.

At Sony, for example, developers follow a simple prescription: "Turn ideas into clear targets." In other words, Sony doesn't tell developers to come up with a portable product. It tells them, as Sony's chief executive did when he launched the Walkman project, to originate a product the size and weight of a paperback book. That's how Sony achieves its overall goal of creating new conveniences, methods, and benefits. Similarly, successful pharmaceutical product leaders such as Pfizer and Glaxo target their projects toward developing drugs to treat medical problems for which solutions have thus far been elusive.

Product leaders in high-tech target their R&D toward the development of devices that are smaller, faster, lighter, cooler, cheaper—whatever constitutes better performance—than those existing. They keep their target descriptions simple, because they appreciate the power of a one-page vision. Their work is guided by a lucid picture of the goal shared by everyone in the organization.

Once an enterprise comes up with a breakthrough, the product rarely sells itself. Edison knew that breakthroughs often create products for which there is little initial demand. After all, if the product was barely imaginable, how could anyone know they wanted it? Thus, the cultivation of markets must go hand in glove with breakthrough product development.

Product leaders have to prepare markets and educate potential customers to accept products that never before existed. Otherwise, new products that push the state of the art can arrive too far ahead of their time. The struggling Remington company, for instance, developed its typewriters in 1874. Although Mark Twain bought one immediately and even invested in the company, it took 12 years for the larger market to catch on to this unfamiliar device. The idea for the microwave oven was serendipitously discovered in 1946 by Percy Spencer, who stood in front of a radar magnetron and noticed that a candy bar melted in his pocket. It wasn't until 1967, though, that people started buying microwave ovens, and the products didn't become indispensable staples in American kitchens until the 1980s.

Every breakthrough product develops sales demand at a natural rate. Academics refer to it as the rate of diffusion of innovation, and it follows an S-shaped curve, slow at first, then fast, and then leveling off. The challenge for product leaders is to push the rate of diffusion beyond what is natural and common, to get demand to climb faster, earlier. Larger-than-life launches, early adopter programs, and massive marketing education are all in the repertoire of product leaders.

With his innovative business operating model, Edison achieved an astounding record of research and development productivity. Current thinking on so-called high-performance teams and learning organizations owes a lot to Edison's process innovation.

Running a top-drawer product development process today, however, has become yet more challenging than in Edison's day. Companies now play for much higher stakes. Single entrepreneurs may still work out of their garages, but product leaders are big companies that have to make big commitments and big investments to bring new products to market—even to niche markets. Johnson & Johnson, for instance, invested several hundred million dollars to develop its Vistakon unit's disposable contact lens. Adding to the cost of product development today is the difficulty of maintaining the narrowly specialized research staffs necessary when new product development involves the intricacies of advanced technologies. Adding extra risk to the process is today's huge and ever-expanding knowledge base, which has opened up so many opportunities for innovation and provoked so many ideas at competing companies.

How do product leaders hold onto their position when cost and risk conspire to break their grip success? They manage a portfolio of development activities; they employ structure and process; and they learn how to manage their people.

DIRECTING THE PORTFOLIO OF ACTIVITIES

Product leadership demands that companies place bets—big ones and small. Deciding where to place those bets is the challenge. Where do you put your talent? What projects should people work on? How much money should you allow them to spend? Large numbers of ideas—none of them sure bets—contend for a finite supply of investment dollars. Each project team clamors for attention and makes a compelling case, but no company can fund every idea—or even every project it undertakes.

The most successful product leaders find ways to quickly narrow their portfolios. Product leaders since Edison's day have concentrated their resources on the handful of opportunities with the greatest potential to hit big. They pick their opportunities the way Fidelity Investment's former mutual fund manager Peter Lynch picked stocks. Lynch looked for "10-baggers," stocks that could yield a 10-fold return on their initial investment. One 10-bagger can compensate for a lot of laggards and losers in an investment—or product development—portfolio. It's the big hits that put a shine on the whole portfolio.

Glaxo concentrates its product-development machinery on a narrow band of about four primary therapeutic categories with large markets in the developed world. Anti-ulcerants and respiratory drugs are two of them. The company always seeks to narrow the field and prioritize opportunities because, as Richard Sykes, Glaxo's CEO, says, "There is not safety in numbers, only greater costs." If Glaxo succeeds, it will succeed big.

Product leaders don't just follow their gut feeling. They do everything possible to squeeze as much uncertainty as possible out of their ambiguous undertakings. At the end of the day, however, it comes down to the vision, insight, and judgment of a few people at the top of the organization to set or reset the company's direction.

THE ROLE OF STRUCTURE AND PROCESS

Volumes have been written on how to turn an apparently hapless product development process into a streamlined, replicable, well-structured set of activities. The advice generally suggests that if you take the messiness out of the system and eliminate the dysfunctions, you'll get something that *should* work if you can only ensure that everyone follows the streamlined methodology and adheres strictly to the prescribed steps.

That's what the coaches say. The players know it doesn't work that way—generic operating procedures and structured processes don't produce winning products unless they are designed to play into the drivers of individual behavior. Two of those drivers are a thirst for problem-solving and a distaste for bureaucracy.

Does that mean that structure and process play little or no role in product development? Hardly. Product leaders create flexible organizational structures and robust processes. They allow resources to move toward the most promising opportunities during development and during the resulting product's market life cycle. They are continually shifting resources to the project or market where the action is. "Make hay while the sun shines" is the prevailing attitude in these companies.

Robust processes are those that enable people to flex their muscles and minds without creating disruption. They provide efficient coordination, while accommodating inventiveness and discipline. The product leaders we observed applied several principles to accomplish this fusion.

Principle One: Keep people on track by organizing the work in a series of well-paced challenges, each with a clearly defined outcome and a tight deadline. Intermediate milestones, and the chance they create to celebrate interim victories, generate the excitement on which talented people thrive.

Principle Two: Create business structures that don't oppress. Large companies replicate small ones' entrepreneurial spirit by breaking people up into teams or clusters. Or they locate their research labs in the woods, away from the companies' potentially stifling headquarters. Thermo Electron, the Waltham, Massachusetts-based $1.2 billion high-tech company, has spun off eight divisions into autonomous, publicly-owned companies. They sell everything from perfume patches to pollution-measuring instrumentation and mammography equipment. Although Thermo controls a majority of all of the companies' shares, their CEOs act as entrepreneurs, not divisional caretakers. Seven to 12 percent of shares in each new company are reserved as options for that company's management. Thermo's central management is unable to impose any plan on any company in the Thermo family. The company expects to spin off about 10 more companies in the next decade.

Principle Three: Stress procedure where it pays the biggest dividend. Almost invariably that means emphasizing procedure during the final leg of the product-development effort. Looming drop-dead dates and commitments tense the work. The velocity and frenzy of activity accelerate. Teams burn money and their own energy at hyper-speed. Directions seem to change by the hour, and tempers flare. Every past sin and omission comes back to slow down progress. It's make-or-break time.

Product leaders have learned that they can avoid the embarrassing "oops!" of discovering too late that engineering's design can't be manufactured, that the product can't be serviced, or that it's not what customers want. Their solution is to work cross-functionally and pay close attention to the later development stages. There, they apply themselves to understanding the root causes of glitches. They map their processes and workflows backwards to learn what created those time delays and misdirections. Diligence pays, they say; you have to go back to the source of the problem.

Many times the source is some residue of the traditional stove-pipe system—research passing its handiwork to design, which then passes it to marketing, which sends it on to production planning. Those slow, serial systems are relics of the past. Product leaders have learned that if they cut their development cycle time in half, they can get twice as many shots at targets only half the distance away. As Ed McCracken, Silicon Graphics' CEO, explains it, the later the company starts its development projects, the more it enjoys the benefits of the latest customer input and the newest technologies.

TALENT: PRODUCT LEADERS' FOREMOST RESOURCE

Product leaders never forget the basics—that more than anything, talented people are the agents of the company's success, and that, ultimately, breakthroughs are born of individuals. That's why Paul Cook, founder and chairman of Raychem, says that a huge proportion—maybe 20 percent—of his and his top management team's time is spent on recruiting, interviewing, and training. That's also why Ed McCracken of Silicon Graphics says, "We place bold bets on the people we hire and then give them the freedom, indeed push them, to make bold bets too."

Managing people comes down to finding, motivating, growing, guiding, and keeping talent. Product leaders have to bring together more—and more diverse—people to craft "insanely great" products, as

Apple Computer founder Steven Jobs would call them. Heroes at product leadership companies are people—or increasingly, teams of people—who beat the odds to do something that supposedly couldn't be done.

Time and again, from behind the glitter of breakthrough products, emerge the names of the superstars that guided them to market. Fidelity Investments, a product leader in mutual funds, has the likes of Peter Lynch and Jeff Vinik, both exemplary stock pickers. *The New York Times* and *The Washington Post* have their Pulitzer Prize winners. Drugmaker Glaxo has the two-person team that ignored one failure after another in pursuit of Zantac, the largest selling drug in the world. Sometimes, stories like these assume mythical dimensions and the heroes grow larger than life. They become emblematic of the organization's carefully nurtured culture.

Finding the most original, best, and brightest people may mean encouraging the gadflies, the concept champions, the mavericks, the unconventional, and the eccentric. At Sony, company recruiters scout engineering schools and universities for technical talent. But instead of simply tapping those at the top of the class, the company gives the nod to students who are *neyaka*—which means they demonstrate a combination of humility, creativity, and versatility. The way Sony sees it, outstanding engineers should be able to move easily from one project to another and solve problems not encountered at school.

Product leadership companies stretch people's potential by throwing tough challenges at them and by inciting collegial "rivalry." Great colleagues bring out the best in each other. People ratchet up one another's standards and performance levels. Product leaders impose few constraints on people beyond the momentous goals they've been assigned, thereby allowing them to rise to the occasion. Product leaders do what it takes—whatever it takes—to stimulate employees' imaginations. Do creative employees want technologically lavish offices, fitness gyms, or espresso bars where they can meet and talk? They'll get them, if that's what it takes.

The highest form of recognition—the award that these talented people most treasure—is selection for the next, even more challenging mission. Star players are always anxious to learn what's over the horizon. Sure, stock options drive them harder, and getting extra time to explore absolutely anything, as at 3M, strengthens their loyalty to the corporate cause but the factor that most quenches their thirst for professional gratification is a stream of mind-blowing problems.

That reality creates a challenge for management. It has to assure that gifted but idiosyncratic people are able to work harmoniously in the cross-functional teams that are prerequisites to the creation of manufacturable and marketable products.

EXPLOITING THE VALUE LEADERSHIP ADVANTAGE

By and large the tinkerers, engineers, scientists, and artists that dream up new product concepts don't know how to do the grunt work of marketing and selling them—what we call value exploitation.

In this respect, product leaders are experts. Neither meek nor modest, they know how to get customers to pay a price premium for their high-value products. And they know how to time their new product releases to keep rivals from clipping their margins.

Customers don't pay price premiums unless they perceive the product to be worth the extra cost, which means that product leaders can't set outrageous prices. Neither can they stand on their powerful brand names alone. To customers, brands are nothing more than the product leader's implied promise—the promise to deliver unmatched value. If prices exceed the value of that promise, the product leader's edge disappears.

To manage prices over a product's life cycle, the company has to consider that what was once new and unique will soon begin to lose its sparkle. Its perceived value to customers will decline. So the company has to lower the product's price, but carefully time the price reduction to gain the most long-term profit.

Product leaders are also expert at launching new products, because they understand that success isn't happenstance. A new concept doesn't have a chance in the market unless it's visible, well-understood, and cannily promoted. It's got to stand out from all the other value propositions bombarding customers daily. Product leaders launch their offerings with a big bang. Creating larger-than-life events to introduce new products is their way of ensuring that customers grow hungry with anticipation.

Johnson & Johnson's Vistakon division used a big-bang launch to push its disposable contact lenses into a market that knew nothing of the product. Would you entrust your eyes to a product that would be thrown away after several days? Vistakon had to change customer perceptions. It needed to create and disseminate a reassuring message not only to prospective users, but to ophthalmologists and other eye-care professionals. Vistakon's launch probably cost three times the price of an average product introduction in that marketplace. Overkill? Maybe. But the extra money spent—small compared to R&D costs—got the product off to a solid start.

Disney is a master at shaping customer anticipation. Its big-bang launches, which are particularly effective for fashion items, books, movies, and other one-time novelties, create the perception that success, if not already achieved, is practically certain. Such launches convey the impression that people who don't sign on risk being left out. The Disney movie *The Lion King*, for instance, came onto the market amidst glorious hype. Opening tickets went on sale two months early. Disney screened the movie with film buyers to ensure bookings at the best theaters. Disney's consumer products division kicked in with a massive merchandising campaign, big theater displays, a promotional tie-in with Burger King, and more. The idea was to create a self-fulfilling prophecy of success. It worked; box office receipts from the movie's first weekend set a Disney record, and the movie has gone on to become the third-highest grossing film in history.

Product leaders tend to be proud, protective, and ferociously supportive of their brainchildren, which means that they hate to back off from pricing that matches the scale of their launch. Sometimes, though, they have to. In spite of a big-bang launch, customers may not fully appreciate the value of a product. In such cases, product leaders aren't bashful about initially offering the product at a lower price premium than it deserves. As customer appreciation for the product grows, so will the premium customers are willing to pay. Mercedes Benz followed this logic when it launched its Baby Benz in the United States at less than $20,000. Mercedes is doing it again by initially pricing its new C-class cars at less than $30,000. Likewise, Lexus initially marketed its top-of-the-line cars at a mouth-watering $35,000. Once the product sticks and customers recognize its true value, product leaders can kick prices back up.

Product leaders' support doesn't vanish after the product launch. If anything, these companies grow more aggressive when rivals appear on their turf. In the leaders' minds, the market is theirs—they created it and how dare these upstarts intrude? Product leaders live by the adage, "Give no ground." Although their lifeblood is product breakthroughs, product leaders don't shy away from product extensions—new varieties that spur increased demand, additional segments of customers with somewhat varied needs. Leaders seek out every market niche, but they don't create product extension just for the sake of denying shelf space to competitors. The new versions and flavors they add are surgically aimed at embedding add-on value for specific customers and market segments identified by their bull's-eye marketing efforts. All along, the price premium that product leaders obtain is calibrated to make customers feel great.

Inevitably, some competitor's offering threatens to undermine a product leader's breakthrough. Long before that happens, however, product leaders exploit their value-producing operating model by retiring their own products. In fact, as soon as they commandeer the market with one new product, they begin thinking of how to drive it into obsolescence. They want to be champions and contenders at the same time, which requires them to create another product so attractive that it puts the present one out to pasture.

The high-tech industry abounds with examples of product leaders that follow this rule—Tandem, Hewlett-Packard, and Intel, for example. Intel has practiced it with great skill in the microprocessor industry. "Double machine performance at every price point every year" is the dictum of CEO Andy Grove. Intel always has multiple teams working on subsequent versions of a product. When the 486 chip, the brain of many personal computers, was just entering the market in the spring of 1989, a new team was already developing the concept for the fifth-generation chip, the Pentium. Another new team is already at work on the P-7, the seventh-generation microchip.

While a company's left hand prolongs product life with upgrades, enhancements, and other value-adding features, the right hand builds the next generation. Does this create tension within product leadership companies? You bet! But it's the tension that makes these companies vibrant. It keeps them busy managing the dynamic balance between the defense of existing products and the introduction of new ones; between unbounded creativity and the concerns of fiscal practicality; between getting the product right and getting it to market; between betting on a few big ideas and nurturing a broader range of maybes. It's this tension that defines product leadership companies.

7

ONE COMPANY'S EXPERIENCE– INTEL CORPORATION

CHAPTER 7

ONE COMPANY'S EXPERIENCE– INTEL CORPORATION

Like cats, the microprocessor chips made by Intel Corporation are blessed with nine lives—or so Intel claims in a recent advertisement. In a graphic display, the ad shows Intel's third-generation 386 chip as having exhausted all nine of its lives; it shows the fourth-generation Intel486 processor with two lives left; only the Pentium processor, Intel's newest and fastest microprocessor, still has all nine lives ahead of it. What the ad leaves out is mention of the Intel engineers who have been working since long before the Pentium chip was released to design the successor chip that will put down even this fast cat.

Intel understands that if it doesn't outdo itself in the microprocessor business, someone else will.

"Performance demand is doubling about every 18 months for us and most of our competitors," says John Crawford, who managed the design of the Pentium processor. "You fall off that curve and you get ground up. Just to maintain parity, you have to push constantly on all fronts."

Intel defines the discipline of product leadership. What makes the company an especially good example is that it has had to tune its business to compete in a very competitive pack. Although Intel today supplies microprocessor chips for three of every four personal computers sold, its 75-percent share of the PC market is far from assured. Tough competitors make 90 percent of the remaining microprocessors sold in the world. Other major players, which together with Intel still control less than half the market, include NEC, Motorola, Toshiba, Hitachi, and Texas Instruments.

Along with competitors in microprocessors, Intel faces rivals in multiple market niches. Some of them pursue strategies similar to Intel's. Motorola, which builds the chips that power the Apple Macintosh line of personal computers, joined a consortium with IBM and Apple to design the PowerPC chip that will displace some of its own. Motorola, as a product leader company itself, would love to get its hands on some of Intel's market share.

Yet other competition comes from operationally excellent manufacturers that make chips with performance characteristics similar to Intel's. Cyrix and Advanced Micro Devices, for instance, aim to offer the lowest prices on 486 clones. AMD has a contract for Intel-compatible chips with one of Intel's largest customers, Compaq Computer. Cyrix recently allied itself with IBM to make Intel-compatible chips in IBM's plants. Even Microsoft, whose products fuel the demand for Intel chips, is joining the competition by creating guidelines that will permit other chipmakers to produce microprocessors that will run Microsoft programs.

Intel combats competitors, both other product leaders and operationally excellent firms, not just by designing chips with more features but also by pushing the limits of its manufacturing technology. It squeezes the creations of clever designers onto increasingly smaller pieces of silicon. It also stretches to make chips in ever-higher volumes and with ever-better yields (the number of chips made from each silicon wafer). The initial Pentium chip had Intel stretching—"almost beyond the breaking point," says manufacturing chief Gerry Parker—the limits of chip-making technology.

Intel makes much more than its flagship Pentium product. The Intel line contains other microcomputer components and modules, automobile electronic components, a range of industrial and telecommunications equipment, PC enhancement devices, networking products, and supercomputers. In each case, Intel stretches the limits of technology, both in products and in processes. That is the company's chief strategic means of remaining a product leader. The company spends lavishly on R&D—42 percent of its 1993 net income—to remain on top of multiple markets. By stretching as it does, the company enjoys a big share of the profits from a semiconductor market with annual sales approaching $100 billion worldwide and growing at 20 percent each year.

In the conversation that follows, Intel executives provide a picture of the hands-on way they work, and of the management methods they have developed since Intel's founding in 1968 as a maker of semiconductor memory chips. The conversation also shows that Intel relies on other techniques in addition to virtuoso innovation to retain product leadership. Among them: precisely targeting customers, educating them about complex products with missionary zeal, feeding customer ideas back to marketers and designers, operating with a team-based quick-change organization, constantly trying to go one better than competitors, and disciplining people's innovative ideas to meet the demands of new markets. The conversation helps explain why Intel has made history by producing the world's first microprocessor in 1971, and becoming, in 1992, the largest semiconductor-maker in the world.

Contributing to the Intel story are Dennis Carter, company vice president and director of marketing in the Corporate Marketing Group; Albert Yu, senior vice president for the Microprocessor Products Group; John Crawford, an Intel Fellow, who managed the design of the Pentium processor ("Fellow" is the title reserved for the highest rung on Intel's technical career ladder); Gerry Parker, senior vice president and general manager of Intel's Technology and Manufacturing Group, which encompasses all component manufacturing and technology development; and Kirby Dyess, vice president and director of Human Resources.

Trumping Your Own Success

The urge to innovate, to create breakthrough products, is deep, almost uncontrollable, at Intel. The company shelled out about $1.1 billion on R&D in 1994 and another $2.4 billion in capital spending to deliver on CEO Andy Grove's commitment: to make "the fastest chips in the newest applications." The company takes huge risks, swinging for a home run with each product. When one engineering team chalks up a win, another sets out to knock the legs out from under it with a better product. Birth and death, innovation and obsolescence are part of the daily life at Intel.

> John Crawford: Intel's constant push to innovate is a result of a combination of technical curiosity on our part, encouragement by upper management, and forces of the marketplace. One technology factor is the constant impulse toward

ever-smaller devices, ever-faster devices, larger dies [chip sizes], an ever-increasing transistor budget [number of transistors on a chip], and higher megahertz [chip speed]. It is a fascinating thing. You are on a treadmill. You have to innovate or you will be out of business. If we were doing warmed-over 486s instead of Pentiums, we would be in good shape today, but a year or two down the road the product would start to look pretty passé.

Albert Yu: It is interesting to note that we are the ones who are generating the waves, if you will. Others are not. The waves are the new generation of products, and the Pentium chip is a new wave that is just starting. It could be that we'll be initiating the next-generation wave by 1995 or 1996. We are never satisfied, and we are most critical of ourselves. It is a never-ending drive to do better work, and complacency is the farthest thing from our minds.

Gerry Parker: The Pentium chip was among the most difficult chips that we have ever had to manufacture. Its manufacturing shows how far toward the limit Intel pushes technology. It is a huge chip, at the physical limit of the lithography tools [which put transistors and their connections on silicon]. Because of the chip size and the performance required, we had to stretch the technology further than we liked.

Not that this is anything totally new. When we started the 386, that, too, was at the limit of our capability; and the 486 wasn't much better when we introduced it. Fortunately, though, as painful as it has been, we are able to handle the Pentium, and we are getting good yields in production.

MANAGING THE PORTFOLIO

Sustaining product leadership involves some pain, because no sooner does management bless one portfolio of development projects than market and technology trends accelerate, slow, or sprint off in another direction. The project portfolio, a shrewd and sensible collection of concepts when it was constructed, blows apart. Then, in a fit of reevaluation, managers pit existing projects against each other and against new concepts from outside the company to build a new portfolio.

Dennis Carter: People at the executive staff level pay a lot of attention to outside input. We listen to the field and go out on visits with ears open to hear what is actually happening. We read the roll-up of information as it comes in and discuss it in the group, with the idea of possibly shaping some new directions for the company.

Gerry Parker: Right now the big dilemma is that while we have a strong share of our market segment in CPUs [central processing units] and desktop computers, trends in the world seem to be pointing in a different direction, toward information appliances. Is the TV going to be the next home computer? We don't think so.

Albert Yu: Intel has a lot of flexibility to focus on new directions. For example, we had been in the microprocessor business since the 1970s, but it wasn't the mainstream of the corporation. Andy Grove, our CEO, is the one who decided that the future was in microprocessors. We made a strategic shift from memory to microprocessors in 1985. Right now, Andy's vision is taking us into communications. We are developing a lot of communications capability, and my guess is that in five years or less we will be a major force in computers and communications. Andy always has a tremendous sense of urgency, and his personal drive drives all of us.

I do sometimes review how we allocate our resources, what percent of our total resources goes to what. I'm afraid we don't always put enough into the new areas, by which I mean not adding resources but diverting existing resources. It is easier to stay with the status quo, following a linear progression. Making a right-angle turn can be very painful.

THINKING RIGHT TO LEFT

What so distinguishes product leadership firms from most other companies today is that product leaders don't slavishly follow the voice of the customer. The customer can't define for them the next breakthrough product. Market researchers can't define it, either. Although product leaders like Intel stay close to the market, they listen to the astute comments of customers only to hone their *own* judgment of the future. Customers can help them get the details right.

At Intel, the definition of the product is the first step in its development. The target may be far off, even far out, but engineers know where to direct their creative energies. That's what we mean by thinking right to left—starting with a visionary concept crafted by leading thinkers and then instructing people to fire the arrows of their imagination straight to the target, not just anywhere into the future.

> Gerry Parker: Since we are a product leader, we are always described in the press as being an innovator. That's true, but innovation at Intel needs careful definition. When people talk about classic innovation, I always think of 3M. People at 3M are geared to invent one crazy thing after another. At Intel, the challenge is different. We also have a lot of bright people who want to innovate in totally new directions. But they often don't understand that we have set a course that keeps us ahead of the market in computers and computer communications. We are essentially innovating to solve a very specific problem, not just any problem. We are not a think tank of innovation for its own sake. We are making a product.
>
> To translate an idea into a product, you try to imagine the next logical step in the progression. What features does the market want and what can you put into your product? What will the technology allow you to do? By answering such questions, you basically define a product.
>
> To make sure that manufacturing is preparing for the future, we look at the product and continually update it. We focus on what we think is going to happen. We are trying now to develop a model of what the future PC requires in

total, what kind of components. The crucial element is trying to anticipate the demand five years out, both ours and for the rest of the PC market.

We need innovation, but a directed, disciplined innovation aimed at solving very specific problems. Even with the original microprocessor invention, Ted Hoff was trying to figure out a better way to make a calculator programmable. It was innovation to solve a specific problem we faced.

Albert Yu: One interesting process is trying to foresee what is going to be needed by the time we develop the successor to the Pentium processor. No one knows for sure, of course. We talk about those questions with a lot of software people, customers, end-users, and so forth. From that we have a pretty good idea of what the next chip should be like.

I should note that once we had completed the Pentium processor, we redefined who our customers are. Traditionally, our customers were only those OEMs [original equipment manufacturers] that buy products from us. That is a shallow view. The real customers are the final users of the computers that incorporate our products. As a result, we established a very close link to our customer's customer. That gave us direct contact with our end-users, as well as with our traditional OEM customer. That policy has given us some new ideas from the outside. Our traditional customers give us a tremendous amount of input in terms of the architecture of the product. The end-users give us a lot of input about the more global things that they want to have in the product.

Dennis Carter: Information from end-users has resulted in many product modifications, though seldom of a major or structural kind. We know what we want the architecture of our next-generation microprocessors to look like, but, beyond that, we'd like to know what users will find useful. Feedback gives us options for fine-tuning and for making things fit better.

The field people came up with so much information that the factory people had a hard time evaluating all of it.

Therefore, we put in place an information technology board of advisers, a dozen information technology managers from Fortune 500 companies who meet about two or three times a year. We spend a couple of days with them, presenting product plans in depth and soliciting their feedback. That allows us to get very focused information on things that would normally be under nondisclosure.

It also allows us to bring product-planning people—marketer, engineer, or both—from the factory to meet with real users and get a direct response to our plans. It's a good way of coordinating information from the field, expert evaluation, and our internal views.

Gerry Parker: With everything beginning to be tied in with communications technology, the feeling is that we have to move our horsepower toward meeting that challenge. Coming up with the kinds of products that will be successful in this new area calls for innovation, of course—but again, market-driven innovation. There is a lot of trial and error in that part of the business and more product innovation as well. It tends to be driven by what we perceive the customers want and what it takes to make the computer more useful.

ON THE RUN

Product leadership pivots on technology, of course. So as technology advances at a faster clip, so too must product leaders' pace of innovation. Leaders like Intel have hardened themselves to the chest-pounding rhythm of innovating year in and year out. They have accustomed themselves to the occasional tail winds and downhill stretches that give people a chance to catch their breath in this marathon, but they have also prepared themselves for the long uphill climbs and the final push to the finish. That's life in the endless contest for product leadership.

Dennis Carter: The fast change of our technology drives the pace. We tend to move and think at a tempo unknown in other companies. Products are achieved, go through their life cycles, grow obsolete, and are replaced by another product that has been gestating quietly all the while. The life cycle here is fast.

Albert Yu: If we have a brilliant new product idea, we can turn on a dime. We can have a new product out the door in three months because we are so close to the marketplace. It's not always that fast, because we have our own bureaucracy to ensure a regular and orderly process, but we can also cut through that process when we need to. The development of the IntelDX2 microprocessor, a member of the Intel486 family, was a classic example of putting a project on the fast track.

John Crawford: It's challenging and exciting to work in this environment, particularly with this double-the-performance-every-eighteen-months challenge we're faced with. If we had the same kind of progress in the automotive industry, we would be driving 11,000 miles per hour faster than in 1982, and we'd fly from L.A. to New York in six seconds instead of six hours. This performance treadmill can get stressful at times, so the challenge is to keep from being burned out by it. One thing that helps here at Intel is the satisfaction of knowing that millions of people have liked your product well enough to buy it. The calming effect of that knowledge helps relieve the stress.

Albert Yu: Keeping everyone focused over the long development time of the Pentium processor was a difficult challenge. It was like running a marathon and realizing halfway through that you are very tired. Oddly enough, the most difficult time is not during the process; it is afterward, when you suddenly feel empty. I think that the excitement is so great during the development that all of us are physically tired but mentally exhilarated. People see themselves making history—and it's true. Every member of the team is helping to write new technological history. The tough part of it is when things are not going well. At those times, we have to take some time off and then come back recharged.

John Crawford: One of the leadership roles I contributed to was in the later phases of designing the Pentium. What happens in the last six months of development is the final push to get the thing done even as others are talking about the next generation. We had launched a team in Oregon to build a more advanced microprocessor, and they were coming up with ideas of what they could do with their transistor budget. They were making good progress toward another leap in functionality and performance from that product, and a lot of the new research was already bearing fruit for the next generation. The shine of the Pentium processor was beginning to pale in comparison to the gleam in the engineers' eyes for the generations to come. So I helped keep our team focused in getting the Pentium product to market.

Albert Yu: Designing the chip is actually the simpler part of creating a new microprocessor; the harder part is to ensure that it runs on the software that runs the previous-generation processor, and the check for compatibility turns out to be a bigger task than the design effort itself. The technical challenge is enormous. The task is similar to that of software development, in which coding is a very small part of the problem—it is the testing to make sure there are no bugs that is the main software concern.

John Crawford: In a commercial enterprise, you want to take calculated risks to get a shrewd balance between the size of the risk and the benefits of the results. If you take risks on all fronts, you are dead. At my level, we take risks to get more out of our transistor budget, or to shorten time to market. Greater risk usually translates into larger design effort.

Some of the risks we take are to be more productive. We could, for instance, take a chance that would help cut our time to market. A good many of our important risk decisions involve using a new computer technique that promises extra performance. We can quantify the benefits as much as possible by saying that, if we succeed, it will improve the performance on this suite of applications by, say, 15 percent. But there will be costs: The die size will be 20 percent larger; it will take more design time and thus X more hours to complete the program. That means we go to market that much later. So is it worth it?

Dennis Carter: From my personal experience, I feel encouraged to take risks. I have never felt punished for a failure that occurred simply because I took a calculated chance against the odds. If it was poorly executed, however, that's another story—then I must accept the blame. The Intel Inside campaign was extremely risky. We spent a lot of money and laid ourselves open to a possible negative reaction. Moving from print advertising, which we had done for years, to TV, which we had never done, was risky. Picking a small Salt Lake City ad agency with limited TV ad experience was a big risk—and some people let me know it.

HIGH-PERFORMANCE TEAMS
THAT CROSS ALL BOUNDARIES

The risks at Intel get spread around. That's not to say that nobody there is accountable. The reverse is true. Every individual on a variety of cross-functional teams makes a commitment to the success of the project. Intel has become particularly seasoned in guiding the progress of large teams, the kind necessary to figure out how to fit millions of transistors into less than a thimbleful of space. Such complexity demands coordination of an unprecedented kind, one of the biggest demands imposed on a product leader like Intel.

Albert Yu: The trick is to pull together as a total corporation—to look at product, manufacturing, promotion, sales, and advertising as one entity. Compare it to an orchestra made up of different instruments but all performing one symphony together. Every player has a major role. If we are out of sync, we are in trouble.

Gerry Parker: There are a lot of cross-functional teams at Intel; we have found that if we get the teams working across functions, we get better progress. They will learn from each other. We value shared learning highly.

Dennis Carter: Much of what we do is accomplished by ad hoc teams. Different parts of the company face different challenges. The technical teams, like those on the development of a long-term project such as the Pentium processor, are of course different from those in marketing. Speaking for marketing alone, I can say that people find the teamwork invigorating. It is a consulting environment, where the program may change from week to week. Our people like being involved, and they seem to relish unexpected challenges. The team formation (and then its dissolution) is not an issue because there is always something new to move on to.

Albert Yu: I was the general manager of the microprocessor business at the time the Pentium processor was developed. We worked in a single team, all in one place. The drawback was that managing a team of 100 is quite a challenge. It took nearly as much innovation to manage the development team smoothly as it did to invent the technology.

We knew from the outset that small teams tend to do well and large teams don't do well because of communication problems. To deal with this, we broke the chip into small blocks, each handled by a subgroup. Within that smaller group, communication was good. The challenges arose when interteam communication was called for, and so we tried to draw some clear guidelines as to how teams would talk to one another.

We had to be flexible about the team approach, so if one group fell behind, we would give it added support. A town meeting format basically accomplished that purpose. We had to do some shuffling around and move people to different tasks in order to make sure all pieces of the mechanism would catch up with one another. We made such decisions weekly, and if needed, daily.

John Crawford: What we had, essentially, was a set of overlapping teams. Each individual working on the project really belonged to a couple of different teams. The challenge for management was to organize in accordance with natural team structures even though the makeup of a lot of the natural teams was contrary to standard organizational boundaries. For example, in the chip design team, we had logic design specialists and circuit design specialists. The logic people worked at the more abstract level, while the circuit people got intimate with the electrons. A team of logic and circuit specialists would implement a certain piece of the chip, both working on the same geographical area.

Because the logic and circuit people had different specialties, the bureaucratic tendency would be to keep them separate, in two different departments. But that would set up an organizational barrier to the natural team format. So to encourage those informal teams, we had a joint management structure with a circuit manager and logic manager and a group of specialists reporting to them. If there were any negatives to this structure, they lay in too much emphasis on the geographic teams. It hindered our development of the methodology. We had a little fracturing of the design styles that we had to counteract. The trick is to get the right balance.

Gerry Parker: We also have teams assigned to process-improvement projects. For example, we have teams at our factories abroad that are working to improve cycle time. One of the problems we had was the lag time in shipping wafers from Israel to Manila to the U.S. and then to a warehouse or a customer. That cycle used to take about thirteen days. We had a team in each of those factories looking into why it took so long, and they worked the time down to four days. Every day we saved last year meant 75,000 more Intel486 chips produced—our daily output. Elsewhere in manufacturing, we have joint engineering teams that are aimed at improving the output of the equipment and improving yields.

FLUID ORGANIZATION STRUCTURE

Because people, above all, provide the basis for a product leader's success, Intel takes care not to shackle talented people in one corner or another of its organization. Instead, managers move valuable people around the corporation to tackle challenges that promise to generate the most value. Managers, Intel executives concede, are the toughest to move around, but move they do, since Intel runs a pure meritocracy: People's stature and role depend on their delivering the value that assures the company's success.

Dennis Carter: The company is designed for rapid change. We reorganize frequently, and people get shuffled around a lot. Internal organizations are created, flourish, then get dismembered. People are used to that, though, and they don't build up long-term identities with any one group. Our identities are simply with Intel.

Kirby Dyess: Changes in our business drive redeployment of human resources. We are constantly moving valuable resources from lower areas of return to higher areas of return, to functions that provide higher value to our customers.

When we started formal redeployment four years ago, we found that one of the toughest groups of people to redeploy was managers. They were used to their own organization. They were used to the structure that they had managed.

One manager who was redeployed formerly had 80 to 100 people reporting to him. The need for his function went away as our business model changed. Thus, his organization wasn't adding value, and yet it was adding costs. The result was that the employees were redeployed into other positions at Intel. The last person to be redeployed was the manager, who had actively helped place his people.

We didn't just ignore this manager. He had done an excellent job in previous positions and now had effectively planned and executed the dismantling of his own organization and job. He was given an opportunity to start up a new organization within a different business, a major challenge, and he was successful.

Albert Yu: The policy of moving people from one area to another is a good one because once people achieve some small amount of success at a new task, their confidence builds. The frequent transfer process has turned out to be much smoother than we originally imagined. There is the psychological problem of moving from the satisfactions of a completed project to the uncertainties of a new one. But if people really believe the future vision, they will be able to move forward.

Dennis Carter: As for the downside, some people view the environment as chaotic, with no structure they can hang on to. They may like what they actually do, but they don't understand their role in the system—and this even applies to some managers. It is certainly true that roles are ill-defined, but our feeling is that carefully defined roles lead to a rigid and static organization. Some people are comfortable with that and some are not. On the other hand, the tasks are not ill-defined. You know precisely what your goal is on a given program or project. People who can live with a certain amount of uncertainty and yet never doubt their ultimate goal fit in well at Intel.

Kirby Dyess: We are not against demoting people at Intel. There are times when managers or employees get into positions where they are way over their heads. We confront the performance shortfall. The message isn't, "Okay, you're out the door." Instead, we decide that, since the manager has performed well in other positions but isn't working in this new one, we'll bring them back a grade level. Often that means the person reports to someone who was previously reporting to them.

Managers have to come to grips with their own egos in these situations. We have moved some managers into different roles, somtimes back one, two, or three grade levels. Once they gain success, they often move back up the organization; the experience turns out to be a valuable one. The organization as a whole is very accepting of demotions and changes. I think the reason is because we don't have a lot of focus at Intel on grades and titles.

COACHING THE KNOWLEDGE ACROSS THE COMPANY

How does Intel transfer not just people but intelligence across the company? Mainly by coaching.

Kirby Dyess: As people move around to new positions, we find that a lot of coaching is helpful. I'm a good example. I was doing start-up businesses prior to coming to this job, with as many as 50 people and as few as two or three. In this position, I have a huge organization.

For coaching, I have been able to use my predecessor, my boss, and a couple of people on my staff.

Coaching goes on throughout the company. One of the ways it crops up is during subordinate input into reviews for managers. We ask subordinates questions such as: Is your manager modeling Intel values? Is he/she adding value to you and your team? This is valuable input for managers.

We also ask the employees to coach managers on their professional development plans—and actually evaluate their managers periodically on that plan. The managers essentially stands up in front of his subordinates and acknowledges what the employees feel are strengths, weaknesses, and areas for improvement. In a recent survey, 50 percent of managers with development plans involved their subordinates in the plan. The subordinates were part of the coaching mechanism to make the manager and the organization more effective.

There is a lot more time and attention that goes into the selection process today at Intel. There are more people involved in the interview loop, including peers, team members, and subordinates.

Before, the hiring manager was looking for particular skills. Now, in addition to having the skills, we need someone who can interface well with other people, because they are likely to work in a team. We also need someone who can thrive in the Intel environment where we are constantly moving resources.

We used to define a job in terms of responsibilities, years of experience, and degrees held. Now we are including the expected deliverables the person must provide in the first 12 months. In this way, we can better identify and prioritize skills and experiences the candidate must have to get the results required. When you start with only the classic job description, oftentimes you hire somebody that really can't deliver the expected results.

EXPLOITING THE PRODUCT ADVANTAGE

Even market leaders' new products won't sell if people don't understand them. In launching its third-generation microprocessor, the Intel386, the company found people hesitant to switch from the earlier 286. Potential buyers didn't see what the successor chip offered that was new. So the company took to the streets to educate customers. It also found that appealing just to its direct buyers—the manufacturers of personal computers—wasn't enough; it had to talk directly to the end-users. Today, Intel is still talking to users. It targets different audiences with different messages. The result: much faster sales.

> Dennis Carter: We are a product leader, so our marketing has two imperatives. First, we must create an understanding of the technology and how it fits into the marketplace as a whole. We have to make clear to customers the benefits of the new technology over the old. Second, we have to explain who Intel is and what we stand for. Successful as we have been, we can never take our credibility for granted.
>
> In general, we try to build marketing programs around some kind of long-term view. The best example of that is our Intel Inside marketing program which defines those enduring

"who" and "what" foundations of the company. The fundamental strategy of Intel Inside is to build an awareness of Intel and the role we play.

We've been advertising more and more to end-users since we launched the 386. And about the time the market shifted to the 386, we had to start selling the 486. It became very complex. That was when we realized that we needed to communicate something about what Intel stands for, our identity, and not just the product we were selling at the moment. So we started playing around with ideas that later turned into the Intel Inside program.

Whenever we launch a new product, we must also spell out its benefits so end-users will understand the new architecture, its attributes, and where it excels. However, this tactical, short-term marketing must mesh with and reinforce our long-term goals of reinforcing Intel's identity.

Our largest consumer audience is made up of those who buy PCs for individual, mainly home, use. They have another level of interest in the technology. They get a lot of their information at the retailers where they buy the product; thus, being able to reach them with information at the point of sale is important. Some of our field sales force has been redeployed to call on resellers and supply them with educational information. We changed our audiences' level of information acceptance. We succeeded in developing their expertise in listening and absorbing.

Above all, our message is that we produce the best and the newest. Today we are the absolute world leader in our product field.

THE DISCIPLINE OF PRODUCT LEADERSHIP

As veterans of Intel make clear, the company follows no simple formula to achieve product leadership. Multiple competencies, extending well beyond outstanding product design and manufacturing, have contributed to the company's earning its preeminent position in its industry.

What most helps Intel deftly ride the currents that swirl through its industry? Workers, managers, and indeed an entire organization of people who are poised to change focus and direction in a snap. People throughout the company who can tackle tasks in a multidisciplinary fashion, often working near strangers on temporary teams. A self-correcting system in which all employees are dedicated to feeding back information on the company's internal and external performance. Designers who can maintain their focus on precise market need. And a marketing and sales organization that understands that customers need continual reminding of the value of buying machines with Intel inside.

It's a symphony, as Albert Yu says. No wonder the company is a leader in advanced technology in the field of PC microprocessors. And no wonder it is creating the architecture for tomorrow's—and the day after's—computer industry.

> Gerry Parker: I would submit that when we had to have some marketing brilliance we got going on the Intel Inside program. When we had to learn how to manufacture advanced microprocessors, we got on with it. I suspect the right thing to do is to define your market and know exactly what you must do to be successful in it. You adjust your available resources to produce what will satisfy that market. The one who stays ahead of it and reacts to its changes almost before they happen is going to be the successful one, as opposed to the company that is just super-good at what they are already doing.

8

THE DISCIPLINE
OF CUSTOMER
INTIMACY

CHAPTER 8

THE DISCIPLINE OF CUSTOMER INTIMACY

Computer salespeople in the 1970s and 1980s, losing many a close sale to venerable IBM, often gave but one excuse: "Nobody ever got fired for buying from IBM." Corporate America, it seems, was so convinced of IBM's technological mastery that computer chiefs—fearing a hearing with the boss if anything went wrong—dared not buy from any company with a less formidable reputation than IBM.

The salespeople that lost out to IBM, however, indulged in false thinking. What they should have understood was that computer chiefs were not so much fearful of reprisal as they were happily dependent on the depth and range of services that IBM, and IBM alone, delivered.

IBM CEO Thomas Watson, and later his CEO son, Thomas Jr., had built that Rock-of-Gibraltar reputation by directing the gaze of his people to but one master: the customer. His unparalleled success in making IBM the source of expert advice on the use of business machines hobbled competitors for years. Although his company entered the computer industry later than many, it was soon to achieve dominance in the fastest-growing market on earth.

Think of IBM at the height of its market leadership in the 1970s. What did it stand for? The best products? Not in many people's minds. Although its products performed solidly, they usually did not come with the benefit of the latest innovations. For the up-to-the-second advances, customers had to turn to IBM competitors, which pioneered on-line operating systems, virtual memory, minicomputers, and most other breakthrough capabilities. IBM was an adopter of these innovations, but not often the leader. Did IBM offer the best price? Hardly. It left that to other companies that knocked off its designs. IBM's unmatched value was not in product or price, but in the extraordinary level of service, guidance, expertise, and hand-holding it offered clients rocked by the turbulent changes in information technology.

To the data-processing chiefs at many companies, IBM was a total solution. It offered methods for planning new business applications, for training development staff, and for managing data. It helped these chiefs explain to senior management the need for a steadily increasing budget. It pitched in when a system wasn't working, diagnosing the problem and getting it back on track. It took care of planning which new machines were needed, which had to be upgraded, and how to integrate new technology. It sent clients off for periodic education to deepen their technical knowledge and broaden their managerial skills. In the 1970s, you might say the data-processing chief's job in many companies was well near impossible without the support of IBM.

Thomas Watson's success was built on offering an unmatched total solution in the marketplace, delivered through an operating model tuned completely to that purpose. His focus and dedication have been emulated by many customer-intimate market leaders today. Of course, more recently IBM has lost its way. It has gotten a black eye from customers that believe it has been over-exploiting its longstanding client relationships. It moved too cautiously toward newer concepts of computing, in part because it had not maintained parity in product innovation. Rather than mining for untapped potential, it became stuck trying to exploit its old solutions to clients' old and new problems. This is hardly a position from which to maintain customer intimacy. The Rock of Gibraltar that was once IBM has turned, probably forever, to shale. Its glorious history and painful decline offer important lessons in both "how to" and "how not to" succeed as a customer-intimate company.

When we talk of companies offering the best total solution, we mean that although they don't necessarily offer the lowest price or the latest product features, they still provide a better overall result for their clients than anyone else. They do that by attending to a much broader range of their clients' needs. The lowest price or the best product isn't the best value if the client doesn't know how to use the product effectively, fails to fix its "broken" management processes, or lacks the skill to achieve optimal results. Customer-intimate companies are unmatched at addressing all of these client limitations.

The most common characteristic of customer-intimate companies is that they offer a unique range of superior services, from education to hands-on help, so that customers can get the most out of their products. Home Depot and Nordstrom, for instance, continually surprise customers with knowledgeable, caring, and unhurried sales personnel who give sage advice about the products and their application, whether it's replacing a broken light switch or assembling a wardrobe. Their competitive advantage is found in their people, who leave customers wondering: "Why can't other retailers offer me the same level of personalized attention?"

Customer-intimate companies personalize basic service and even customize products to meet unique customer needs. Airborne Express, for instance, continually surprises its customers with tailored solutions that competitors can't match. Exceptionally early delivery, same-day courier service, special handling of windshields and vaccines. Although such services are often out of the question for their competitors, they are routine for Airborne. Similarly, Cable & Wireless, the Great Britain-based telecommunications company, offers its clients customized networks and service features that couldn't be obtained from MCI or British Telecom. Its customers marvel: "How can such a large company craft a solution that meets my unusual needs?" For a customer-intimate company, it's all in a day's work.

Other stars of this discipline go even further, helping revamp the client's business processes that involve the use of their product. Milwaukee, Wisconsin-based Johnson Controls, for example, provides its building management clients with experts in energy use, who offer plans that can change building design and management. A.G. Edwards, the stock brokerage firm, trains its brokers not to push particular securities, but to work with its clients to clarify life objectives (put kids through college, care for parents, enjoy comfortable retirement), construct an achievable plan, and develop in the client a discipline to stay the course. Cott Corporation has achieved spectacular success in the private-label soda business not solely by offering inexpensive soda, but by partnering with supermarket retailers to change the basic processes by which private-label products are managed, marketed, and sold.

Another element that customer-intimate companies stress is a willingness to take on responsibility for achieving results. They will often put themselves at risk to further their clients' success. They will even take full responsibility for an operation, and deliver a guaranteed result. Roadway Logistics Systems, for example, has taken over all responsibility for inbound logistics at two Ford assembly plants. Hundreds of component suppliers, 14 transportation companies, and all warehouses are managed by Roadway Logistics to achieve just-in-time delivery. Ford gets the result without any need to transform warehousing, scheduling, or freight management. These are now completely Roadway Logistics' problems.

Similarly, Baxter International's hospital supply unit is continuing to take over the job of managing supplies for its customers. For a fixed fee per patient-day, Baxter now provides all the basic supplies to a hospital and takes responsibility for managing inventory and logistics right down to the nursing station and the supply cabinet. Baxter also takes responsibility for shrinkage. The system is up and operating at over 400 hospitals already. Baxter's customer-intimate approach is sure to keep shaking things up in the hospital-supply distribution business.

Some of these companies have taken the art of customer intimacy even farther. While companies such as IBM have focused on helping clients better use products, many companies have found other components of best total solution at which to excel. Toronto-based Cott Corporation, which makes private-label soft drinks, was asked by Safeway Supermarkets for help in marketing its store-brand line of soda pop. Cott saw that Safeway needed to do more than simply revive sluggish cola sales—it needed to improve the performance of all its private-label products. Cott suggested alternative approaches to package design, merchandising, promotion, and production formulation. Cott even showed Safeway how its idle bottling plant capacity could be used to make product for other Cott clients.

In true customer-intimate fashion, Cott was able to look beyond a simple syrup-and-water solution—beyond a standard offering which would have been enough to satisfy Safeway—and instead offer a total solution to Safeway's needs. Safeway and Cott both derive greater value from the resulting relationship.

Cott's customized solution extends to marketing. Rather than follow the lead of Coke and Pepsi, which pour hundreds of millions of dollars into brand advertising, Cott is extremely selective with its promotions. It spent just $2 million to launch the Safeway program, but it also enlisted the help of the people who could be the most effective in selling the cola: Safeway employees. Cott reminded them that private-label soda helps preserve Safeway's profit margins—and their jobs. The cola Cott gave the Safeway workers convinced them that it tasted good— good enough to recommend. In effect, Cott created 76,000 new salespeople for Safeway Select Soda and achieved a stunning 35 percent market share within four months of introduction. Working similarly with Wal-Mart, Cott achieved and has maintained a nearly 50 percent market share for Sam's American Choice cola against product leaders Coke and Pepsi.

THE OPERATING MODEL OF CUSTOMER INTIMACY

Successful customer-intimate companies are those that have become expert at their customers' business and at creating solutions. But no matter what their formula for combining help in using the product, advice on transforming the underlying processes, and responsibility for achieving results, these companies deploy an operating model based on design principles very much like those followed by Thomas Watson.

Watson's business was highly client-driven. It was proactive, change-oriented, and proud of its superior knowledge of the application of its products. IBM measured its success by its clients' success. Complex solutions were crafted and integrated by sophisticated account teams and dozens of specialized service and product support groups. The result: a total solution tailored and coordinated by a dedicated account team and efficiently delivered through selected specialized experts. Current thinking on matrix- and team-based management owes a large debt to Thomas Watson.

Watson knew that he didn't have a chance without deep knowledge of his clients. He built the best-trained sales and service force that the world had ever seen. He aggressively invested in new practices, new ideas, and new approaches to the application and management of business machines. The result was that IBM's thinking about the best use of its products was generally two steps ahead of its clients. IBM could attack the root causes of a problem while the client was still focused on the symptoms.

Deep customer knowledge and breakthrough insights about the client's underlying processes are the backbone of every customer-intimate organization today. It's not by accident that Cott is investing heavily in new approaches to supermarket category management; that Roadway Logistics has hired and developed savvy logistics talent to rival any consulting firm; that Baxter International knows more about the use and management of supplies in a hospital than any of its clients; or that Zeppelin, the German distributor of Caterpillar earthmoving equipment, knows more about the effective management of heavy equipment than a highway construction company could possibly hope to. Every one of these companies has built a body of expertise in account teams and specialized service groups that is the foundation for customer-intimate relationships.

But becoming and staying customer intimate requires more than building client knowledge and having expertise in reengineering client business processes. Watson believed that to be effective one must offer more than just service. He maintained a very broad product line that was configurable to the specific needs of a client. His products may not always have had the latest features, but he knew that an average product tailored to a client's very specific needs is often better than the more advanced, but inflexible, product. Organizations like Watson's are not obsessed by the leading edge; they embrace solid, tested products that can be tailored to fit clients' needs like a glove. They produce unmatched value for clients who don't necessarily want the very latest product—just the best result and help in obtaining it.

Since customer-intimate organizations mold themselves to their customers' needs, an extraordinary variety of activity is always percolating across a broad set of accounts. This creates a laboratory for developing better application support, for discovering new process improvements, and for testing new kinds of relationships where responsibility increasingly shifts from client to supplier.

Watson knew that a decentralized organization better achieved tight relationships with clients. He understood the importance of empowerment and the critical role of individual initiative. He summed it up in a one-word motto—THINK—and disciplined his sales, support, and service forces to live the motto every day in their interactions with customers.

As IBM grew—and grew!—Watson maintained a rigid focus on account control. His management system was based on detailed measurement of account penetration. He focused his control system not on profitability or market share, but on share of customers' spending. Every customer-intimate company knows that the critical objective is share of client. The worst failure in a customer-intimate organization isn't to lose money; it is to lose a client. Client by client, Thomas Watson set targets for penetration, development, and growth. As far back as 1924, he hired R.L. Polk Company to use its city maps, directories, and other proprietary data to find specific potential customer accounts. Polk helped to identify the character and potential of more than two million of them.

To operate this type of control system, one needs specific, detailed, and integrated customer data. Operationally excellent companies may be able to tell you a lot about their transactions, but customer-intimate ones have computer systems that can tell you a lot about their customers. This is especially important when it's not one person, but a team of people that serve an account.

For more than 30 years, IBM's competitors were endlessly frustrated by Watson's success. Many of them tried to adopt his approach, but failed because they thought IBM's success stemmed from the sales and service force. These wanna-bes let product development and manufacturing continue on their course, while their field forces labored mightily to become intimate. It didn't work. Customer intimacy, as Watson built it and others have discovered, requires the efforts of the whole company. Sure, the banner carriers are sales and service people, but without the complete alignment of product development, manufacturing, administrative functions, and senior management, one can't achieve a total solution for customers.

Consider, for example, product development. Customer-intimate companies don't sell products at the leading edge. Their business depends on a stream of products that represent evolutionary improvement, not revolutionary change. That's because they've piled on top of their product lines layer upon layer of services to address clients' limitations in using the products. Some of these services are significantly destabilized, even destroyed, by breakthroughs in product. So companies like IBM have preferred what their customer-intimate clients preferred—steady, controlled, incremental evolution of product coupled with expertise that leads the clients through changes in their application and management.

Many competitors, with more quickly evolving and cutting-edge products than IBM, failed to emulate IBM's customer intimacy. Digital Equipment Corporation, during its years of ascendancy, hired countless IBM sales representatives in an effort to duplicate their success. Despite large investments, Digital failed at its effort. Many of the new employees complained bitterly about the difficulty in receiving support from service and product units that were not passionately client-driven. Without the full backing of the rest of Digital, the new salespeople could, at best, offer customer responsiveness—not intimacy. Working diligently, they might have been able to satisfy their customers' expectations, but they were unable to guide them, to change their ways, and to build strong interdependent relationships.

Watson's principles of customer intimacy have been applied today in industries as diverse as retailing, distribution, industrial manufacturing, consumer packaged goods, and logistics. Roadway Logistics is decentralized, client-driven, and change-oriented. Its account teams are highly skilled, knowledgeable in the client's business, and actively developing new approaches at the leading edge of logistics management. The company has built a core process for delivering a total solution that integrates a diverse and deep set of specialized services at the point of contact with the account. It has become a model for other parts of the transportation industry. Roadway Logistics has discipline, the value discipline of customer intimacy. Like Cott, Nordstrom, Airborne, and Zeppelin, Roadway Logistics knows its share of each customer, knows its account profitability, and understands the lifetime value of each client.

The formula for success of customer-intimate companies has gradually changed over the years. More customers today are concentrating on the parts of their operating model that are critical to their own success. They are looking for partners to take responsibility for secondary processes, to outsource and deliver results, and to increase their flexibility. Knowledge about business is becoming more and more specialized, leading to greater reliance on outside advisers. Logistics, marketing, and information technology, for instance, are all areas in which expertise has become deeper, more specialized, and ever-changing. The new customer-intimate market leaders have had to expand and adjust Thomas Watson's principles to fit a modern world of enlightened employees, hollow delivery systems, and ever-deeper customer relationships. Let's look at each area in turn.

THE MANAGEMENT OF PEOPLE

The central management challenge in customer-intimate companies is to assemble, integrate, and retain talented people who can stay at the forefront of new paradigms and techniques that affect their clients' business. The most sought-after employee has tremendous skill at effecting change within client organizations.

Good ideas today are cheap, a dime a dozen in our real-time, inter-networked, fast-paced world. Brilliant concepts and practices are disseminated with stunning speed. Today's business magazines are so on top of new developments that they now describe the latest innovations long before the Harvard Business School can get around to writing a case study about them. Competitive benchmarking and best-practice studies have become standard elements in most organizations.

What's still in short supply, though, is the ability to effect change, to get things implemented, to make things happen. That's the value provided by customer-intimate companies. In this way, they operate much like management consultants, who know that the proof of their value is found only in results. Deeply rooted in the culture of a customer-intimate company is the sense that if the client does well, I've done well and we've done well. The most cherished award at a customer-intimate company: a prize from the client, recognizing that the company has been an instrumental part of its client's success.

Stories abound in these companies about employees who have gone above and beyond the call of duty for their clients. An example is the story about the Four Seasons Hotel doorman who found the briefcase of a guest who had already checked out. Assuming that it contained important papers, the doorman rushed to the airport, caught the next air shuttle, and delivered the briefcase to the forgetful fellow. Heroic? Well, yes. More important, though, the story adds to the mythology that typifies the way the hotel runs. The doorman is now an icon, not just an isolated character in a crazy story. The message to employees: Four Seasons Hotel customers deserve nothing less than service that dazzles, that awes.

Are the many stories of employee heroics that issue from companies such as Four Seasons, Home Depot, and Nordstrom literally true? Perhaps not. Perhaps they've benefited from embellishment. The point is that the mythology supports a strong culture, one that tells employees: Do whatever it takes to please the customer.

Just as in management consulting, however, results for the client have started to feel in recent years as if they're getting harder to come by. Why? Because customers are more sophisticated than ever. That requires a sophisticated response. At Roadway Logistics, customers aren't assigned salespeople. Instead, "directors of logistics development" study the customer's operations, evaluate its needs, and determine what value Roadway Logistics can bring to the arrangement. These logistics directors don't close a deal and disappear. They stay close to the process long after the operations people move in. One prime example is GM's Lordstown, Ohio, facility, where Roadway keeps 50 managers and warehousing people on staff at all times.

This symbiotic relationship between Roadway and its customers plays out in other ways, too. One of its customers, John Deere's Horicon, Wisconsin, tractor-manufacturing facility, was facing stringent state packaging-disposal laws that took effect in 1995. Deere decided to start returning containers to suppliers, for which it needed a computer system to control and reship the containers. Roadway wrote a software program to meet that need and then proposed a system to manage the flow of all materials, not just containers. Today, the company manages not only the flow of materials and containers from all John Deere suppliers but also plans all of Deere's transportation. It is even installing equipment to wash and repair Deere's lawn-tractor containers. Such new business stems, at least in part, from Roadway asking every on-site employee to look for ways to increase its penetration into the customer's business, a hallmark of nearly every customer-intimate company.

Customer-intimate companies, more than product-leadership companies and far more than operationally excellent companies, tend to resemble a loose collection of people who somehow all deal with a set of customer-driven issues. What you won't see are clones—employees who all walk, talk, and think alike. Watson's view of the company man has been supplanted by a new style for a new era. Customer-intimate companies need a broad set of skills and styles to get the job done. Their employees are adaptable, flexible, and multitalented, allowing them to deliver just about any reasonable—and sometimes, unreasonable—response. This means that the person with the right background or skills has to be willing to jump in when needed—even if what's needed at the moment is not exactly his or her job.

Customer-intimate companies hire a mixture of seasoned and inventive people. They need the depth of insight that years of work within the client industry brings, but they also need irreverent, out-of-the-box, transformational thinkers, because the world is changing so rapidly. This blend of experience and inventiveness prevents skills from becoming either obsolete or irrelevant.

Home Depot is widely recognized for its mix of people. It is the only large building-products retailer to place experienced tradesmen—carpenters, plumbers, electricians—on the floor to help its customers. Cott Corporation has hired many of its people from the supermarket industry and teamed them with bright, aggressive, young business school types.

Customer-intimate organizations also use their clients to stay at the edge of new thinking. Studying these companies provides a giant learning laboratory. Like consulting firms, they practice a form of Robin Hood egalitarianism: Rob from clients rich in insights and give to the poor. Every customer-intimate company has developed techniques for sharing among account teams the general insights on best practices gained in working with a particular client. This institutionalization of knowledge is a key to their competitive edge.

Sometimes no boundaries seem to separate the customer-intimate company from its customers. It's hard, if not impossible, for an observer to tell where one company begins and the other leaves off. When Roadway Logistics extends its relationship with a manufacturer to the point that it not only provides logistics management but also performs component assembly and delivers the assembled components just-in-time to the client production line, where's the border separating one company from the other?

With so much activity directed toward clients' individual needs, it is easy to imagine a customer-intimate organization pulling apart, flying in all the directions that its clients are heading. What keeps it together? What is at the center that can hold? Well, for one, the mechanism for sharing learning among account teams ensures that most teams remain dependent upon the organization for a lot of their new insights. All employees working in accounts recognize that much of their success rides on the powerful service groups that stand behind them. When these account representatives get hired by their clients, as sometimes they do, their effectiveness is usually seriously diminished. They have lost access to these shared resources; they have lost the leverage of being an outsider; and they have lost the learning that comes from being in an organization that deals with dozens, if not hundreds, of similar client situations.

HOLLOW DELIVERY SYSTEMS

Many customer-intimate organizations, like Roadway Logistics, offer a staggering range of products and services to their clients. How do they do it? How do they amass such capabilities and make them available to their account teams? The key for many of them is that they "rent" rather than own many of these capabilities. Many customer-intimate companies are hollow businesses.

The strength of these companies lies not in what they own, but in what they know and how they coordinate expertise to deliver solutions. Cott is a perfect example. It uses its knowledge of soft drinks to design and implement sophisticated private-label branding strategies for customers like Wal-Mart and Safeway. It sells these retailers an awful lot of soft drinks, but it doesn't make the concentrate that gives the product its flavor, and it doesn't bottle the product. In fact, Cott, one of the fastest-growing beverage companies in the world, doesn't own a single bottling plant.

To achieve production flexibility, Cott assumes a general contractor's role, designing a total solution for a retailer's private-label needs and taking responsibility for the solution's execution. It relies on RC Cola for the concentrate and a network of bottlers for the product. When it comes to the design of the label and package, Cott relies on The Watt Group, a design firm that it controls, but is maintained as an independent entity. Cott coordinates and integrates many functions with subcontractors to create a profitable retail brand product, which is what its customers want.

A major success factor for many customer-intimate companies is their network of product and service capabilities. It is a network under virtual control of, but often not owned by, the company. This approach has two clear advantages. First, the company is able to broaden the range of its total solution by extending its network into areas in which it lacks capabilities. Second, it can avail itself and its client of components that have other value propositions of lowest cost or best product. For example, Cott Corporation may not be able to produce a soda at the lowest cost, but it can contract for it from an operationally excellent bottler.

IBM's failure to broaden its network of capabilities beyond what it created and managed internally exacerbated the recent swoon in its fortunes. IBM approached the customer with an attitude that IBM, and IBM alone, was going to serve them. By closing its portfolio of capabilities to outside developments, IBM cut itself and its clients off from a huge array of capabilities. As IBM found out, customer-intimate companies today can't afford the not-invented-here syndrome.

One might wonder how a customer-intimate company can extract a profit from reselling other companies' products or services. It can't. Repackaged offerings from other suppliers offer little value. But if the company brings a combination of subcontracted components and its own services—advice, reengineering changes, responsibility for results—plenty of value remains on the table from which to extract a profit.

CREATING DEEP RELATIONSHIPS

Customer-intimate companies take the long view. Thus, initial transactions with a client don't have to make financial sense by themselves, so long as the long-term relationship promises to be profitable. These companies are more than happy to make investments in building relationships, but to receive an eventual return on their investment, they have to retain their clients. A steady client is a lasting asset; a one-time client is a poor investment.

So they avoid or shed clients that don't have deep relationship potential. Thus, the customer-intimate company must be able to distinguish the mirage from the real, and it must be willing to walk away from business that might generate only short-term revenues.

Customer-intimate companies steer clear of pure transactions. It hurts their business to serve clients that already know what to buy and are shopping only for price (as with airline seats) or product features (as with entertainment products). If they don't require advice and expertise, transaction customers won't find the customer-intimate company's offering particularly compelling, and a customer-intimate company that pursues such customers finds itself competing—and not well—with operationally excellent and product-leadership companies on their own turfs.

To be worthy of a customer-intimate company's attention, clients must meet the selection criteria. Three dimensions of fit are considered. The first is attitude. Is the potential client inclined to see and appreciate an opportunity for joint gain from an ongoing association? Both supplier and customer must see the opportunity. The customer must be open to a relationship in which some independence is lost. If this thought is so foreign that it violates a basic principle, then little opportunity for a relationship exists. The best potential clients are those who feel something is missing from traditional supplier relationships and have a gnawing sense that a greater opportunity exists out there somewhere, but simply haven't been able to define it yet.

The second dimension of fit is operational. The ideal operational fit exists when compelling expertise meets wanton client incompetence. It's hard to be customer-intimate with a client that knows too much. Ideally, the customer-intimate company has demonstrable competence in one of the customer's vital process areas. The level of expertise is such that it overcomes any reluctance to enter into a dependent relationship.

The operating ideal would be the application of the customer-intimate company's superior knowledge to the client's mission-critical processes, because it creates the most value for both sides. The challenge then becomes staying ahead of a client's competency in that process so that both sides continue to reap proportional benefits. But, of course, few clients have any reason to be incompetent in a mission-critical process. As a result, most customer-intimate companies offer total solutions to secondary client processes.

The third dimension of fit is financial. For customer-intimate companies, the ideal financial fit occurs when the customer has large, untapped potential. Cott Corporation, for example, saw huge potential in one of its supermarket clients that was losing money selling private-label soda. The client wasn't aware that it was losing money, because of its poor accounting systems. Nor did it understand the earnings that were possible if the products were better managed. Cott saw this and structured an innovative deal that tapped that million-dollar potential for the supermarket—and handsomely rewarded Cott as well.

Customer-intimate companies continually deepen their clients' dependence on them. One job well done begets another as the client's confidence grows. The company uses this closeness to increase its understanding of the client's business. The process builds upon itself. Eventually, the degree of intimacy between the supplier and the client can move all the way from arm's length to tight embrace. The stronger the relationship, the better the opportunity for a total solution.

Customer-intimate companies typically deepen and broaden their areas of client support to further the relationship. Initially, they extend their current competencies—freight service at Roadway, for instance, and private-label expertise at Cott. The initial engagement gives them access and builds legitimacy. This access allows them to observe and understand how their expertise touches the initial customer need and the process associated with it. At this point, they search laterally for other customer processes where they can add value. Roadway Logistics moved from transportation management into inventory management; Cott moved from private-label supply to category management; IBM moved from developing accounting software for clients to advising them on the management of the accounting process. In this way, the customer-intimate organization positions itself to be ready to respond to the client's next set of problems—perhaps even before the client itself becomes cognizant of them.

EXPLOITING THE VALUE LEADERSHIP ADVANTAGE

Customer-intimate companies create for their clients an unmatched value proposition of best total solution. But how do they benefit from the value they create for customers? There are two answers—growth within accounts and growth of accounts. Where other market leaders might raise prices to exploit their product advantage or grow asset utilization to increase their cost advantage, customer-intimate companies turn yet again to their accounts for their success. Customer intimacy is very much a shared journey. Thus, it isn't surprising that these companies exploit their advantages in ways that also benefit their clients.

Customer-intimate companies make money by tapping the unrealized potential in the client's operation. They share in the rewards with the client, either by directly billing for their value-adding services, or more typically, by charging clients a little more for the supplied products than an operational excellence supplier, perhaps even more than a product leader would charge. In return, they deliver much greater overall results.

Over time, the client appropriates much of the expertise that was once the unique domain of the customer-intimate company. Meanwhile, competitors figure out how to replicate the customer-intimate company's solution and deliver it for an efficient price. When this happens, the client understandably starts to view the original solution as a commodity, and expects to pay accordingly. But customer-intimate companies generally aren't capable of making much money from their services at commodity prices. So they have to employ one of two strategies: Either transfer the capabilities to the client, making the client self-sufficient, or subcontract the work to an efficient supplier that can price it like a commodity.

The point is that, in time, margins shrink on services and products that once commanded premium prices. What can the customer-intimate company do? It must search for new areas of mutual cooperation, new untapped potential within the client organization. It seems hard to imagine that one could continue to do this, but unrealized potential is everywhere. One just needs the visual acuity that comes from intimate customer analysis.

Thus steady progress and growth within each customer account, and in new accounts, is necessary if the company is to continue to exploit its market leadership advantage.

For example, as Cott Corporation's relationships with some of its customers have matured, the company has begun to look beyond the clients' private-label soda needs. They now supply private-label products in several other categories, such as salty snacks, yogurt, and pet foods. They have also been asked by a group of clients to build broader expertise in the management not just of private-label products, but of entire categories of products. These new veins of unmined potential offer opportunities for years to come. The key is to continue to find new client potential, build expertise by leveraging learning across accounts, tap the potential at rewarding prices, and move on to new opportunity when the value is largely realized.

The second way that customer-intimate companies exploit their value leadership advantage is by finding new clients to whom they can bring their expertise. The new client is able to tap years of learning, a richness of experience, and a depth of insight. Existing clients who extoll the virtues of their relationships are often the most productive source of new clients.

SOLUTION! SOLUTION! SOLUTION!

Many customer-responsive companies wonder why they just don't "get it." They complain they're doing everything possible to cosy up to clients, lavishing them with attention and service, and still their results are less than stellar. What separates the mighty from the might-bes? Once again, it comes down to hard choices. For a company to become truly customer intimate, it must decide—and throw its full weight behind that decision—to offer clients: expertise that drives client performance; a willingness to share in client's risks; and real, meaningful tailoring and customization of products and services, not useless "value-added" service. And a customer-intimate company must display the confidence to *charge more*, because it knows it is worth every dime.

All in all, the bright glow cast by customer-intimate companies, what draws to them the most loyal of customers, is generated by a canny weave of strategies, superior personnel with unparalleled know-how, application of the newest and finest techniques to the customer's vital processes, and an extended network of product and service capabilities. That glow signals one thing: solution. Solution, like strategy and reengineering, is a concept that is often referred to, but infrequently practiced. But in the customer-intimate company, solution is the foundation of an aggressive and highly successful enterprise.

9

ONE COMPANY'S EXPERIENCE– AIRBORNE EXPRESS

CHAPTER 9

ONE COMPANY'S EXPERIENCE– AIRBORNE EXPRESS

What company is the fastest-growing express-air carrier in the United States? If you guessed FedEx or UPS, you are wrong. It's Airborne Express. For the last 10 years, Airborne has been growing faster than the industry giant, FedEx. Its revenues have soared 20 percent a year since 1985. Why? Because Airborne offers a unique value proposition to a select set of high-volume customers.

Airborne's proposition is to go beyond providing the industry-standard 10:30 A.M. delivery, or the industry-standard package of logistics services, or the industry-standard tracing of packages. Airborne offers to enter into what amounts to a partnership with its customers—to build an integrated, highly customized logistics and package-delivery service that will improve the business.

That's just the value proposition that appealed to Xerox. To get parts to technicians in time for them to quickly repair copy machines, Xerox needed guaranteed delivery times across the U.S. ranging from 8 A.M. to 9:30 A.M. Upon those early, and necessarily reliable, deliveries rested the success of Xerox's service business.

Airborne came back with a proposal that did guarantee those times. It has since made many more such commitments to assure that Xerox gets parts to the field on the shortest possible notice. It has even coded the beepers of its drivers to signal when each driver is carrying an urgent Xerox package for priority drop-off.

Airborne's value proposition, built on tailored services, special treatment, and partnership, separates the company from legendary market leaders FedEx and UPS. What's remarkable is that by choosing a different value proposition, Airborne has carved out a section of the express-air market, and a base of loyal customers, all to itself.

Airborne has made a name for itself in a market that is fiercely competitive. FedEx, with a 47 percent share, and UPS, with a 22 percent share, dominate Airborne, with 16 percent of the market. To be sure, a rising tide has raised all ships to some extent, as the air-express market has grown steadily over the last decade, owing to such factors as the willingness of shippers to pay a premium for fast, sure delivery, the adoption of just-in-time inventory and production systems, and the demand for overnight small-package delivery in today's economy.

But only the most airtight of ships have survived the rigors of competition. UPS prompted a shakeout of air-express companies by waging a sustained price war throughout the latter half of the 1980s, using its scale and low-cost ground delivery business to hold prices down. That sent average revenue per overnight-air shipment to the bottom of the ocean. Standards of service, however, rose rapidly, with FedEx promising no-hassle service and "absolutely, positively" delivering on time. Companies that wanted to continue to sail in this market had to train and perform like America's Cup racers.

FedEx and UPS have responded by offering a narrow set of services to a broad range of clients. They have leveraged their own value proposition: operational excellence. They have shied from tailoring services and have instead tried to maintain their margins through low variety and high efficiency.

To be sure, all carriers have jumped like circus barkers to tell the world that they spare no effort in satisfying customers, but only Airborne gets intimate with its customers. Airborne can afford to, since one of its strategies has been to focus on high-volume corporate accounts that give it economies of scale. Airborne saves money, for example, by making far fewer pick ups. It then tailors solutions that yield mutual gain. For one industrial catalog-ordering business, Airborne actually acts as the entire shipping department, enabling the customer to take orders as late as 1:00 A.M. Eastern time for packages delivered later that morning.

Airborne can also deliver more intimate service because its work force isn't burdened with serving a multitude of small-fry customers. Along with acting as a third party to manage logistics and inventory, one service Airborne supplies to big customers is the LIBRA automated shipment processing system. Customers installing LIBRA can produce an invoice, weigh and rate a package, and route the package—all with a few keystrokes. No more waiting for the delivery truck.

Overall, Airborne sells customers a package that includes customized basic service, value-added services, and expertise in logistics redesign—all as a partner willing to sit down and talk about whatever arrangement would make the customer more successful. A disciplined approach with those components has helped Airborne grow in a crowded field. "Airborne is totally consistent in the message it gives to employees," says Frank Steele, senior vice president of Sales. "All our managers get together once a year and go through the drill again, to remind everyone of the company's agenda."

Airborne was the result of the merger of two small West coast freight-forwarding companies. It changed direction in 1980 when it expanded into the relatively new overnight air-express industry. It operates its own airline, ABX, Inc.; owns a hub airport at Wilmington, Ohio; forwards freight to 200 countries overseas; and runs a global communications system that gives customers real-time 24-hour-a-day access to information about their shipments.

Conversations with Airborne managers, along with customers that work with them, show how Airborne has established an intimacy with its customers that pays dividends over and over. It shows that Airborne, from the sale, to the service, to the after-service customer support, offers its customers total solutions that competitors have a hard time matching.

The people telling how Airborne made that happen include four managers from Airborne, and two customers. The managers: Ray Berry, vice president of Field Services Administration; Mike Dari, district sales manager for Airborne's Long Island, New York district; Joe Devore, national account manager for Xerox; and Frank Steele, senior vice president of Sales. The Airborne customers: Ken Bram, president of National Parts Depot, a $4 million distributor of printer and computer parts; and Nora Phelps, transportation manager for Xerox's Eastern Distribution Operations.

THE CHOICE AND
CULTIVATION OF CUSTOMERS

Airborne's approach is the antithesis of mass marketing. The company neither courts a mass of customers, nor can it serve well a mass of customers. Airborne seeks out customers with whom it can create a mutually profitable relationship. Once that relationship solidifies, the company cultivates it further for both its own and its customer's benefit. Airborne remains choosy for some simple reasons:

Ray Berry: There's an advantage in our being selective about the customers we serve and the services we offer. The customer needs we have targeted to fill are what we are best at. If, for example, we had large mail-order customers requiring nothing but residential delivery, we might not be able to serve them as well as we know how to serve IBM or Xerox. Since we can't be all things to all people, we pick our kind of customer deliberately.

Joe Devore: Although our company doesn't try to be all things to all people, we provide a premium service in transportation logistics to customers for an economical price, and we work with them to achieve their goals. But we don't, for example, handle residential deliveries for catalog companies like L.L. Bean.

Frank Steele: In our sales effort, we don't have the kind of luxury UPS and FedEx have. Between them, they've spent billions of dollars on television advertising over the years. FedEx, which was one of the first to advertise, and is a very good company, has already established a generic name for itself. There isn't enough money on the planet for us to counter that—so we don't advertise at all. As a result, it's unusual for someone to call us cold and ask to do business: We have to find our customers.

Although we haven't tried to compete in terms of name recognition, we have grown faster than FedEx for 10 straight years without using any public message. It has all been face-to-face selling—and we sell 95 percent of our business directly, maybe even 98 percent or 99 percent.

How does my 300-person sales organization find the high-volume users that best match our mode of business? Years and years of experience have given our sales representatives a pretty good handle on our marketplace. But identifying potential new customers is also an ongoing process of sorting through leads from customer-service people, drivers, and delivery manifests, along with our knowledge of those industries that tend to be high-volume users.

Maintaining partnerships calls for adapting the sales organization to give customers the attention they demand. Airborne has gradually changed the way it operates to care for its prized accounts:

Frank Steele: In 1988, we faced a critical problem in continuing to offer the kind of partnership approach we desired. With only 300 people in sales and with a huge base of business, we were reaching the point where all our sales people were stretched in terms of their ability to manage big-customer relationships. Large customers, particularly multi-location customers, have a lot of complicated requirements and needs, and it's difficult to pay them the proper amount of attention unless you have enough manpower. So we created the position called national account managers. We have 10 of them now. And we will have 11 next year.

Joe Devore: For such companies as Xerox and IBM, which use our services on a national level, we need someone with an overview, someone who can pull together all the information on these accounts. This manager's function has to cross operational boundaries.

Frank Steele: These managers are customer-specific. Each national account manager is responsible for a few individual accounts that total about $20 million of business. That could be two or three accounts. I would say that on average a national account manager handles four customers. Their job is to be seen not as representatives of a vendor but as internal consultants to the customer. They make sure that customers get everything they need from our service. Those managers also insulate us from competitive incursions, and expand our relationships with the customer companies.

The companies that advertise have a broad base of infrequent users, while our customer base is almost entirely made up of frequent users. So that is where we focus our attention, and the national account manager is the close liaison between Airborne and its array of frequent-user customers.

THE VALUE PROPOSITION: TAILORED SERVICE

As Airborne establishes a partnership with a customer, it stays away from peddling the same commodity as its bigger brethren. It customizes service. That customization begins with a superior job of understanding the customer's specific needs. Airborne seeks to eliminate customers' downtime, thereby accelerating their cycle times and saving them money. Differentiating itself in this way, as an agent of its customer's success, has proved to be brilliant strategy in a crowded market.

Now listen to two people from Airborne describe their approach to the client even for basic services:

> Frank Steele: We tailor our service to fit the customer's mode of operation. We agree to provide many specializations that our competitors won't even discuss with their customers. We try to become an extension of each customer's business operation. By paying close attention to customer needs, we've developed a kind of flexibility that goes a long way toward attracting customers and building their loyalty.

> Mike Dari: Our biggest strength in our relationships is an understanding of how the customer company operates and the details of its needs. We try to make sure that our depart-

ments—sales, operations, and customer service—comprehend exactly what customers do and how they do it. To make that work at Luxottica, we met with the customer service, MIS, shipping, operations, and billing departments and got an understanding of what each needed.

Our competitors approach a prospective customer with the attitude of "Here's our program rate—why don't you go with it?" Our approach is more like, "We're offering you a kind of partnership that can do both of us good, which means we can help you become a faster-service organization." We try to suggest different levels of service the customers might not think of themselves.

Now listen to Airborne's customers. Their comments mirror those from Airborne insiders. The services from Airborne improve the customer's business:

> Ken Bram: Airborne was even helpful in practical ways during our transition to them. Their people came to instruct my staff, and they went into our shipping department to see that we understood the new computer setup. I had a slight problem with the software that prints out the Airborne COD tags, so they had my software modified at their expense. Now everything goes out Airborne unless the customer requests UPS.

> Nora Phelps: Xerox selected Airborne through a bid process in 1988. At that time, the industry standard for next-day delivery was 10:30 A.M., as set by FedEx. We had a critical need to get emergency parts to our technicians even earlier. These technicians service our machines in the field. Airborne responded to this and many of our other unique needs with very innovative solutions. They improved the overall service greatly.

How do Airborne insiders create that great service? Listen to Airborne's national account manager for Xerox:

Joe Devore: During the original contract bid process in 1988, we did our best to learn what Xerox needed in relation to what we could give them. From the outset, they were open in explaining to us what their goals were, and how their internal systems work and what kind of transportation they wanted. Xerox had this unique need for immediate delivery of parts of copier machines to their technical reps. They needed especially early service for the reps awaiting those parts at customer locations. In essence, we were being asked to cut the downtime for Xerox copier customers as drastically as we could.

We were presented with a list by Xerox of the locations that requested earliest delivery, and we told Xerox the time at which we could promise delivery in each of those places. Knowing the problem with this kind of exactitude made the answers easier for us. The result is that we have delivery commitments anywhere from 8:00 A.M. to 9:30 A.M., depending on the location. Those commitments are taken very seriously in Airborne and are a visible standard within the corporation. In fact, our district operations managers' performances are measured by that standard daily, and the record becomes part of their review criteria.

To meet those early delivery times, we had to customize our basic services to Xerox's needs. For example, the bulk of Xerox's shipping comes in and goes out of Rochester. We obviously want to provide the earliest inbound time to Webster, a suburb of Rochester where the company has a large facility with 20 to 30 buildings. To save time, we created a specific sort code for the place. When freight comes into Rochester, our people look for containers called C-containers, marked specifically with that Xerox code, pull them immediately, break the freight down to truckload, and get the trucks headed out to Webster.

We have loaded the Xerox early-delivery system with all kinds of readily-accessible information. To track a package throughout the process of delivery, we have coded every scanner of every driver around the country to beep and read

Xerox when there is a Xerox shipment to deliver. The drivers, realizing they have an urgent package, will then prioritize their load plan to make that delivery on their first stop.

We do everything we can to make our service the best. A lot of this has to do with working with Xerox. If Xerox had not been willing to work with us, we would never have gotten to that point. This again is a good example of how the two companies work together.

THE VALUE PROPOSITION: A BROADER RANGE OF SERVICES

Airborne builds on its basic, tailored service by both delivering extra services competitors don't offer and helping its customers redesign the way they operate. One tool that Airborne often leverages is information technology, integrating its information systems with its customers' to improve shipment tracking and billing. Airborne, in essence, constructs an operational structure that reaches directly into the customer's business. Some outsiders would have trouble figuring out where Airborne's business leaves off and its customers' businesses begin.

Frank Steele: Logistics service is changing the face of business today. Elimination of warehouse facilities means that customers can cut down on their real-estate liabilities and distribute their product from our centralized locations— using our faster capability. We now even have major customers asking us to take over their trucking operations—and thus we've become involved in that kind of transportation. This is a typical result of our national account management process, which now includes about 40 customers. We find ourselves taking on challenges we'd never foreseen. But if we can do it, we will do it.

Joe Devore: Among some other subsidiaries that operate in conjunction with Airborne is Advanced Logistic Services (ALS), our third-party warehousing logistics service that operates a truck hub network. Another is Sky Courier, which is our wholly-owned, next-flight-out, immediate-priority shipment service. We have used both of those groups in work for Xerox.

Sky Courier handles emergency orders and follows a rule of next flight out (NFO) on commercial air lines. One of the unusual advantages we offer Xerox is the same rates in New York or Los Angeles—or any other location. We also developed jointly a program called Rapid Deployment. Sometimes a Xerox rep at a customer site can't pick up his or her part in time. A ground messenger from Sky Courier will then collect it and take it to the customer site. That cycle time is anywhere from one to four hours, depending on the market and the service request. This service is currently running in Los Angeles and the San Franciso Bay area, and we hope to expand it into Chicago.

But Airborne not only bends over backward to help customers with its extra services, it bends forward to examine the details of its customers' business, helping customers redesign that business if requested:

> Frank Steele: The closer our sales managers get to the customer, the more needs they see. And the more time they spend in close contact with the customer's business, the more ways we find to satisfy that customer. We find good new solutions for all sorts of operational problems or inefficiencies they may have.

> Joe Devore: In our work with Xerox, we have helped them to find all sorts of ways to save time and money. For example, the field tech reps, who repair the copying machines, used to take the part that needed repair back to the DPC [district parts center]. Someone keypunched information about it into the system, and the part was sent to a repair site to be retooled or disposed of. The turnaround for that process was averaging close to 30 days. We revised the process. We now pick up the parts at the DPC and line haul them to Wilmington. We keypunch the data into the Xerox system through their terminal in our facility and then forward the

parts to the appropriate Xerox location. The result is that we have cut their cycle time to five or six days from 30. We have also reduced their costs, allowing them to reduce overhead, cut personnel involved in keypunching, and speed up their repair cycle.

In another instance, we felt that we could serve the DPCs daily through our truck hub network, which would cost less than air transportation. Xerox had considered this but they weren't sure about its affordability. So to explore this, we formed a joint team of three people from each company. I was included along with a field person and the director of our regional truck hubs. Xerox included an inventory specialist, regional transportation manager, and a transportation coordinator from corporate.

In Airborne Express we also have a division called the business analysis group. The job of the BAG, as we call it, is to look at our costs and the processes by which we do business. We offered to have the director or vice president of BAG work with people at Xerox to show them how we run some of our internal systems. Xerox is a quality-minded company, and to be able to go to them and say that we have a department specifically oriented to this subject and we would love to share its information with you, is a strong means of building our partnership.

Nora Phelps: Many of the shipping problems we bump into stem from our super-fast, and precise, delivery requirements. One of them concerned our use of Sky Courier. Our technicians have a 5:00 P.M. cutoff for our emergency orders. Our cutoff for that-night shipments with Airborne is 6:00 P.M. But a technician might phone at 6:30 P.M. with a request for a part by the next morning. So for that part to reach its destination in time, it had to be shipped NFO through Sky Courier. The technician at the other end would then either have the part picked up at the airport or delivered to the customer. The cost for NFO, however, is about 10 times that of an ordinary Airborne shipment. We were concerned about overuse of NFO—surprisingly, so was

Airborne, since what was high cost for us was also low profit margin for them. Thus, we both had a stake in reducing the number of our NFO orders.

One result is that the national accounts manager for Sky is working with some of our folks in Rochester to reduce the problem. In Virginia, we are trying a few experiments locally. Airborne now sends us a "chase truck" at 7:30 P.M. for pickup of all emergency orders that come in between 5:00 P.M. and 7:30 P.M. That truck goes straight out to Dulles airport to catch the plane that is also carrying our 6:00 P.M. pick-up. The chase truck might carry only a small consignment, but any items that can be converted to overnight Airborne from NFO Sky Courier add up to big savings. This innovation was purely an Airborne idea. We are now, in effect, a partner with Airborne.

BEYOND COMMUNICATION TO ACCESS

Every company today claims a star role in listening to its customers. But Airborne doesn't hear the customer voice from afar like so many companies. Airborne employees circulate among customers, hobnobbing directly with the people that pay the bills. Not just the front-line sales people talk face-to-face with customers. The executives pulling the strings from behind the scenes solicit the thoughts of their benefactors as well:

> Frank Steele: I should point out that being close to the customer means being always accessible for the customer. Someone at Luxottica or Xerox can call anyone at Airborne, because our policy is that all the executives answer their own phones. We encourage customers to call us. We have a very flat organization, with four layers of management between the sales representative and the president.
>
> Anyone can call the president. That openness keeps us alert and responsive. With limited layers of management and emphasis on trying to simplify the decision-making process, we manage to keep our lines of communication open.

Mike Dari: Our company culture has a lot of listening built into it. And that is the message that we try to bring to our customers. Unlike our bigger competitors, such as UPS or FedEx, we have far fewer restrictions on how to do things. We can react much more quickly. I think most customer companies these days want a supplier to work with them and offer a service above and beyond what they are getting right now. That is the reason people come to Airborne.

We try to meet with our customers frequently, popping in from time to time to make sure things are okay. Our operations manager has monthly meetings with the account to find out how things are going. We actively try to receive as much feedback as we can.

Frank Steele: Such face-to-face methods of doing business require goodwill and diplomacy. We never position ourselves as "mining for new business"; that would make for a bad relationship. Our approach is to make sure customers are getting the solutions they want—new profits will come from that.

Our close relationships with customers can often lead to expanded business for us, and not just at large accounts like Xerox. Customers often begin to take advantage of our logistics arm to store their products in our hub warehousing, or "stock exchange," for quick release. Or they use Sky Courier for next-flight-out service.

Joe Devore: At Xerox, our lines of communication are not just at the national account manager level. They operate in relationships between local district operations managers and Xerox district parts center managers. The essence of what we offer is complete communication and understanding about each other and what we are trying to accomplish as companies. It makes us work better together.

The people at Xerox think along the same lines. The one-on-one attention from Airborne helps them in their business:

Nora Phelps: Airborne assigned Joe Devore to manage our account, with excellent results. Joe is invaluable at solving problems, and I keep in constant touch with him. Our ability to communicate was facilitated by an invitation by Joe to view Airborne's operations. Joe thought it worthwhile for me to come out and see Airborne's sorting facility in Wilmington, and I found it impressive. The tour starts at midnight and ends at 4:30 A.M., but amazingly, I was never tired. Once I understood how their sorting process worked, I could much more easily visualize how they planned to work with us.

Joe goes out to meet with our local folks and he brings along the local Airborne folks. He especially talks with our inventory operations managers, who are responsible for the district parts centers. We still have a few glitches, and in the winter months we don't get performance that's as good as at other times, but our field personnel feel that some things can't be helped because of the weather.

Nowadays, the perception is that Airborne people are doing all that they can to give us superior service. Airborne has shown us their integrity. I work out of our regional distribution center, from which we feed parts and supplies to the DPCs. We have a local Airborne representative there, and additional local reps deal with our personnel at the DPCs. The relationship is very comfortable. In all this, Joe is our resource and I have no need to talk with any higher levels in Airborne. That's what a partnership is all about.

MANAGING PEOPLE
FOR CUSTOMER INTIMACY

What particularly distinguishes Airborne from other companies is that its object of attention is not simply the transaction, but the customer. Management eyes not just the hard figures on ledgers but the soft comments from customers in measuring performance. The company has deemphasized budgets alone as the policing mechanism to keep the troops in line.

By opting to put the customer's view, rather than just the accountant's, under the magnifying glass, Airborne has had to adopt a different management style in recent years, as it has refined its customer-intimate strategy.

Ray Berry: Before Airborne restructured in the 1980s, managers were more oriented toward the process, counting numbers of transactions per day, more oriented toward quantitative things. We now have tried to shift the focus to qualitative. Thus, today when there are mistakes we contact the customers to get their reactions about what went wrong and what went right.

Under the old system, when things went wrong, we just let them go and if the wheels came off and the customer complained, we'd just stick the wheels back on without wondering how they had happened to spin off in the first place. We weren't proactive. We did not solicit customer views on how well we did our job.

Today, managers no longer focus on how many transactions were accomplished. They focus on what the customer's perspective is and how well we accomplish those transactions.

Just as the new system required employee changes, it required a different kind of manager. For the most part we tried to pick managers that were very customer oriented. We told them: "Don't look at yourself and think how well you are doing, look at your customers and ask them what more can you do for them."

Mike Dari: We hire the kind of person who will demonstrate that message to the consumer. The drivers, for instance, are the lifeblood of any organization like ours. Our drivers know the importance of the account they deal with. I think that the driver and the shipper are the ones who build the bond between the companies on the most fundamental level. That's why we try hard to hire folks who have good customer-relations skills.

Frank Steele: Our sales people need a set of skills different from those of a standard sales representative because they also act as transportation consultants. To find the kind of flexible people we need, we work hard at hiring. It's interesting that we have the smallest sales organization in the industry. FedEx has almost three times more sales representatives than we do. So it is very important that the people we hire are highly motivated.

INTIMACY FOR THE LONG TERM

Creating deep and strong customer relationships will remain the means by which Airborne grabs market share wherever its rivals slip. The customer-intimate approach has let Airborne skirt the trap of competing as just one more commodity supplier. By sticking to its discipline—defined by premium service, customized deals, value-added services, easy communication, operational integration, regular feedback, and flexible partnerships—customers see Airborne as one of a kind, not one of a group.

Frank Steele: Looking to the future, I can honestly say that I don't see any downside ahead. If we keep on doing the right things locally, then customers will always want to do business with us. The essence of a service company is exactly what the word says—service. The better we do that, the faster we grow and the more successful we are. As long as we don't get sidetracked into kinds of business that we aren't very good at or ones that are only marginally profitable—as long as we stick to what we do well—we are going to flourish.

Joe Devore: What makes Airborne's relationship with Xerox work so well is the fact that neither company is looking at the short term. Both are looking down the road. This shows up in Xerox's bid for continued contracts with Airborne. Xerox's last bid was for three years with a possible two-year extension. When you get to the point where a purchasing group is comfortable with you, they don't get as hung up on annual bidding. In other words, we've established a comfort in the relationship, but one that is based on performance and not on complacency.

Customers at Xerox and National Parts Depot appreciate that lack of complacency:

> Nora Phelps: I see Airborne working with us for a long time, given their continued good performance and competitive pricing. A lot of times when we put out a bid we are checking to see if our current supplier's rates are realistic, but we are not going to end a good working relationship just to save a nickel. As we change and reorganize, I feel that Airborne will continue to be involved in discussions about how they can respond to our new set of needs. The Airborne partnership has been a very satisfactory one for us and I don't foresee any change in it.

> Ken Bram: It's the overall picture at National Parts Depot that I am concerned with, and that means the general quality of service we provide to our customers. The people who represent Airborne are definitely a notch or two above everyone else that I have dealt with in their business, and I'm extremely satisfied with them. To me, Airborne has proven its worth. I think our connection with Airborne is going to last a long time. I have made a major commitment to them, and they have done the same to us.

1 0

SETTING YOUR VALUE DISCIPLINE AGENDA

CHAPTER 10

SETTING YOUR VALUE DISCIPLINE AGENDA

The battle by companies for market leadership never ends. Neither the market laggards, trying to catch up, nor the leaders, constantly challenged by new ascendants, can hope for a cease-fire. Managers from both kinds of businesses lie awake nights—or should lie awake—trying to answer the same question: "How do we compete and win in our marketplace?"

It is surprising how many executives today stumble in trying to answer that question. The longer their answer, the less likely their firm really knows how it competes.

The best answer any company can give is one that defines precisely the exceptional value the company offers customers, and that describes an operating model capable of delivering this value with a fair return to shareholders. Every firm can achieve this, but doing so requires three rounds of disciplined assessment and deliberation.

Before starting, the senior management team should recognize that past habits bias its quest for new insights. Keeping an open mind, and ensuring that the deliberations don't end up as yet another fruitless exercise, requires the team to be alert to the following:

- The tendency for executives to delegate work too quickly after just an initial pass.
- The all-too-common tendency to underestimate competitors that look different or operate in a different way.
- The tendency to pursue multiple markets and value propositions through one business unit, rather than through multiple units, each geared to different markets and to delivering different kinds of value.

■ The tendency to "be team players"—to get agreement among themselves without digging deep enough to bring out and resolve (or at least reconcile) differences.

The three rounds of deliberations have the purpose of getting the senior team to struggle with such issues, to align courses of action, and to gain a deeper appreciation for what needs to be done.

In round one, the team must come to an understanding of where the firm currently stands and why. What are the dimensions of value customers care about? Where does the company stand relative to its competitors on each of these dimensions? Where and why does it fall short? Only by having a shared, fresh understanding of the current situation—based on facts—can a management team evaluate its options.

In round two, the company must develop realistic alternative value propositions and operating models. What would customers perceive as *unmatched* value? Could competitors quickly better that value? What kind of operating model would deliver the value proposition at a profit? What changes would the company have to implement?

In round three, the hard choices get made. Management must commit to one value discipline. How would the operating model work? What change initiatives must the company launch? How must the company restructure? How will the company manage these changes and maintain its focus? How will it manage the associated risks?

This new approach to setting strategy and direction responds to the new rules of competition: (1) put unmatched value of one chosen kind in the marketplace; (2) meet threshold standards in other dimensions of value; (3) make the proposition better every year; and (4) build a superior operating model to deliver on the promise. Focus on nothing else.

To illustrate this approach, let's examine the U.S. toy retailing industry today. It's a business that involves most of the elements of robust competition—entrenched market leaders, aggressive aspirants to leadership, and continuing interest from established firms in other retail industries.

The U.S. Toy Retailing Industry

U.S. toy retailers chalk up about $18 billion in sales each year. The five biggest firms own about half the market. The biggest, so-called category killer Toys "R" Us, operates about 600 stores. It controls 22 percent of the market, and its share has grown modestly during the last three years. The next biggest merchant is Wal-Mart, which is fast gaining on the leader and could conceivably overtake Toys "R" Us by as early as 1997. Wal-Mart's toy-market share in 1994 was 16 percent, up from 8 percent only four years earlier.

Next in line are two other mass merchandisers, Target and Kmart, with a growing combined share that stood at about 13 percent at the end of 1994. Kay Bee Toys, which runs almost 1,000 stores in shopping malls, commands about 5 percent of the market. Hot on the heels of the big boys are a number of smaller players, which together heighten the competition in the market to a fever pitch. Disney and Warner Brothers have rapidly opened 300 specialty shops in premier malls throughout America. FAO Schwarz, a formidable name in toys, has also expanded into malls. A host of new players, such as Learningsmith and Zany Brainy, have entered toy retailing with a focus on child development and education. And a number of other retailers, from computer stores to the Nature Company, have found ways to sell in the toy market.

The toy market, in short, is a roiling, competitive pot. Who does what best? Different companies bring different core competencies to the business. Toys "R" Us has built tremendous competence in demand forecasting and physical distribution. Wal-Mart has done the same, and its capabilities are even stronger. Disney and Warner leverage their competence in building and exploiting image franchises—from Mickey Mouse and the Lion King to Bugs Bunny and Porky Pig. Kay Bee Toys knows how to source and market closeout merchandise. The variety of capabilities and strengths in the toy market resembles the variety in many markets. Given the complexity, how do these companies assure that they profitably employ their competencies to convince consumers, ages three to 93, that their dollar is best spent at one store instead of another?

The traditional tools for analyzing any industry—looking at competitive forces and discerning core competencies—may provide some insights, but such tools are hardly sufficient. They do not focus an executive team on the critical drivers that will deliver unmatched value in the marketplace. Our approach to strategy-setting focuses company leaders on the hard decisions they have to make to pursue their company's chosen value discipline.

ONE COMPETITOR

Child World, a chain of 130 stores, declared bankruptcy a few years ago and liquidated its assets. Its premature demise cost shareholders dearly. The company's equity was once worth $180 million. It cost thousands of employees even more—they lost their livelihoods. The story of Child World's decline and fall is an increasingly common one: undistinguished customer value leads to poor financial performance, which leads to loss of shareholder value, and, ultimately, loss of jobs.

Did Child World's story have to end in tragedy? Did it have no options that would make it a market leader? Let's answer those questions by creating a fictional toy retailer—we'll call it Kiddieville—that runs a business similar to Child World's.

A business magazine writer dissecting the business of Kiddieville could sum up the company's situation succinctly: doing it all wrong, still making a living. Kiddieville hasn't defined its value discipline. It follows an apparent strategy of mimicking Toys "R" Us's approach to the business: large, freestanding stores, busy locations, high volume. The problem is that Kiddieville doesn't have Toys "R" Us's scale, buying clout, or distribution efficiency, so it can't match it on price. To thrive, indeed just to survive, Kiddieville needs to establish some unique dimension of unmatched value.

Kiddieville has tried different things. It tried selling disposable diapers as a loss leader, but stopped when it became too expensive. Later it reversed itself when store traffic declined. It has experimented endlessly with pricing and promotion schemes. None has successfully disguised the fact that its biggest competitors have a price advantage. It has tried an upgraded store environment, offering a better display of merchandise, wider aisles, and attractive signage. But this tactic added to costs without adding to sales. Kiddieville has bounced from guardrail to guardrail while trying to find a sustainable strategic highway.

What is the root of Kiddieville's trouble? It isn't a lack of management depth or experience. The top executives draw on a combined 120 years of accomplishment in retailing. No, the snag is elsewhere. It's the result of nearly everybody on the management team reading from a different map. Success for the executive in charge of merchandising lies in a different place from the vice president of store operations. And their success lies in yet a different place from the head of marketing. No wonder they don't know which road to take.

The senior team lacks consensus on what should be unique about the company, on what unmatched value will put it on the road to market success. Everyone wants the business to thrive, but the executives continue to debate about how to make it happen. So far, they've stirred up an impressive cloud of dust while running in place.

Some of the senior team members recall the good old days when management was clearly aligned. The president at the time discouraged dissident views and irreverent questions. He's gone, however, and times have changed. Can Kiddieville management agree on a single value proposition? What implications would this have for their operating model?

PHASE ONE: UNDERSTANDING THE STATUS QUO

Senior management must agree that all the talent on board amounts to nothing unless everyone sees the company's situation the same way. The debate will drag on until everyone occupies common ground. The only way to reach common ground is to find fact-based answers to five very fundamental questions:

- What are the dimensions of value that customers care about?
- For each dimension of value, what proportion of customers focus on it as their primary or dominant decision criterion?
- Which competitors provide the best value in each of these value dimensions?
- How do we measure up against our competition on each dimension of value?
- Why do we fall short of the value leaders in each dimension of value?

A simple but useful device Kiddieville's management team can use to answer the first question is to think of customer value as simply the combination of the costs customers pay and the benefits they receive.

Each of these value contributors applies to both the products Kiddieville sells and the services it offers. Product costs include price and less-than-perfect product reliability—variables that affect the initial and ongoing cost of product ownership. Service costs include mistakes, delays, and inconvenience, because customers pay with both their time and their money.

On the benefits side, value stems from the product—from unique features that deliver superior results—and from the kinds of service benefits provided. Benefits are measured against customers' expectations, so products and services really offer benefits only if they substantially exceed competitors' offerings. Competitive parity, after all, creates a base level for customer expectations.

For Kiddieville, as for any company, the dimensions of value are specific to its industry. In toy retailing, hassle-free service has three principal components: breadth of selection, location convenience, and efficiency of the in-store experience. Each component puts varying demands on customers' time. A superior shopping experience might result from, for instance, personal shopper services for customers who need help selecting age-appropriate merchandise, child-minding for shopping parents, birthday party planning services, unusual toy demonstrations, and a host of other innovations that might provide a distinctive total-solution for shoppers. Such services are critical—they distinguish one toy retailer from another—since customers attribute value to toy retailers in the category of product benefits only to the extent that the retailer offers unique merchandise. Product reliability doesn't mean much in toy retailing, since consumers attribute this component of value to manufacturers. Price, of course, is a crucial dimension of value in toy retailing.

Once Kiddieville's executives identify the dimensions of value and confirm them through customer research, they'll have to assess the position of each major competitor. Customer interviews are useful in this assessment but not essential. The management team can usually short-cut the process by drawing on its own collective knowledge of the competition. Only when executives substantially disagree should more data be gathered.

Discussions among Kiddieville's executive team might generate the following assessment of competitors.

■ Wal-Mart is an operationally excellent competitor, claiming best total cost. It is the clear price leader in toy retailing, and it offers location convenience for the many consumers who regularly shop in its stores. Wal-Mart offers limited selection beyond core toy products, but expands the selection substantially for the Christmas rush.

■ Toys "R" Us is another operationally excellent competitor. It is also a price leader, but it can't equal Wal-Mart's prices. Toys "R" Us is the hands-down leader in breadth of selection. Its freestanding stores occupy locations that aren't convenient for combining toys with other shopping. The actual in-store experience is typically inconvenient—overwhelming store size, indifferent employees, slow check-out during busy periods—and below industry norms. Toys "R" Us bases its claim of best total cost on two dimensions—selection and price.

■ Kay Bee Toys offers location convenience for mall shoppers and low prices on closeout merchandise—the kind of toys people aren't really looking for. Its value discipline is unclear.

■ Disney, Warner Brothers, and most child-development and educational toy stores offer distinctive products not typically found elsewhere. They are classic product leaders, helped by appealing company images.

■ FAO Schwarz, not unlike many thriving mom-and-pop toy stores, delivers a broader, total solution. Its value discipline is a little unclear, but tends toward customer intimacy. FAO Schwarz emphasizes a distinctive shopping experience and offers a fun environment, and a superior sales staff, as well as some exclusive merchandise. It has difficulty meeting threshold standards on price.

With a consensus on the competition, the executive team can turn its attention to evaluating Kiddieville's own position—where the company stands and why. Now is the time to collect the cold facts to complement the perceptions and opinions of the senior team—or, if required, to resolve differences among points of view.

An analysis team that includes knowledgeable people from various parts of the business could be of great help at this point. The team's job would be to survey customers to find out what levels of value they place on the company's products and services. Three types of customers are important: those who purchase from Kiddieville, those who visit Kiddieville stores but rarely buy anything, and those who shop elsewhere. The objective of the survey is to find out which dimensions of value are important to which customers and how Kiddieville ranks in each of these. This information provides an initial assessment of potential market-segment sizes. The team can then examine the internal workings of the operation to understand how those levels of value are attained.

Once Kiddieville's managers get a clear picture of how their company stacks up in each of the value dimensions, they can look into why its performance falls short. How has the operating model broken down? How can Toys "R" Us afford better prices and FAO Schwarz deliver a better shopping experience and both of them earn superior profits?

The evidence suggests that Kiddieville is the value leader in nothing and that at best it is seen as a Toys "R" Us pretender. Kiddieville's managers have to figure out where each of the company's business mechanisms isn't measuring up, and the whole management team should come to a common conclusion about what's out of whack.

Facing up to reality this way takes guts. Self-assessment is taxing work, and working through disagreements and misalignments among colleagues can be emotionally draining. The process is made even more difficult when management has little teaming instinct. Lots of management teams aren't teams at all but committees, with each member representing his or her self-interest. Painful though it may be, executive realignment—or alignment—must take place, because it's a necessary foundation for building a strategy to turn Kiddieville into a market leader.

PHASE TWO: REALISTIC OPTIONS

In phase two, Kiddieville's senior team shifts from agreeing on what the company is to agreeing on what it could be. Management wants to generate some go-forward options. For each dimension of customer value, it must explore the following questions:

- Irrespective of industry, what are the benchmark standards of value performance that will affect customers' expectations? How do firms achieve these standards?

- For value leaders within toy retailing, what will their standards of performance be three years from now?
- How must the operating models of these value leaders be designed to attain those levels of performance?

Based on such inquiries, Kiddieville has to deduce where the industry might be going in the next two or three years and how standards of value leadership might evolve. Kiddieville wants to clear as much haze as possible from its crystal ball so it can see what effect various changes to its own operating model will have on the company's ability to compete with the current and future value leaders.

Once it has done that, the management team can lock itself in a room and brainstorm. What kind of operating model would enable Kiddieville to take the lead in any one of the value dimensions? How can Kiddieville beat the leaders at their own games? The thinking should be out-of-the-box, and good ideas can come from anywhere—including the operations of value leaders in other industries.

What price points would Kiddieville have to meet to establish clear price leadership over Wal-Mart and Toys "R" Us? Some simple analysis can identify these targets. What cost structure would it take to achieve these price points at a profit? How might Kiddieville build such a structure? What, for instance, would Kiddieville have to offer manufacturers to be able to buy merchandise below Wal-Mart's cost? Would Mattel, for instance, be interested in getting preferred, if not exclusive, placement within the stores? Could Kiddieville lower operating costs by using other companies' assets and operations—expanding into non-toy retailers' space in the weeks prior to Christmas, for instance?

Well, maybe not. Maybe, after due consideration, the management team has to decide that on the price dimension Kiddieville has no good options for achieving leadership.

But what about convenience as an unexploited opportunity? A survey of consumers might confirm their disaffection for Toys "R" Us's in-store service and for Wal-Mart's and other mass merchants' thin selections. Should Kiddieville consider some really unconventional ideas? Are there opportunities to redefine toy-shopping convenience? Kiddieville can examine the operating models of convenience leaders in other industries as a starting point for creative discussions.

The management team needs to explore each dimension of customer value in similar fashion. Could Kiddieville develop a franchise around unique merchandise? Could it redefine the toy-retailing concept by offering a total solution for birthdays and holidays, or meeting the gift needs of grandparents, aunts, uncles, and others? In each dimension of value, what would leadership mean? What would it take to establish it? What operating model could profitably deliver the result?

Such brainstorming has two results for Kiddieville. First, it drives home to the management team just how wide the gap is that separates Kiddieville from its market-leading competitors. At the same time, it generates ideas—from the mundane to the ridiculous, from the obvious to the sublime—about how that gap might be closed. Winnowing down the alternatives requires more discussion among the executive team and, probably, more data from the analysis staff.

At the end of this phase of its work, the executive team has a small set of options. Each incorporates a clearly identified value proposition and a sketch of the operating model required to attain it. The executive team can also consider options for which the value proposition is powerful but for which no viable operating model has yet been identified.

This second phase demands freewheeling thinking and the ability to temporarily suspend disbelief, lest innovative ideas be too quickly squashed. It challenges each executive to disentangle his or her thinking from the burden and blinders of expertise. Kiddieville's executives may see themselves as experts in their industry—but the reality is that they have deep knowledge of yesterday's operating models. Each executive needs a dose of useful ignorance—beginner's mind, as a Zen master would have it—to unleash his or her imagination.

The outcome of phase two is a short list of options that pass the management team's scrutiny. These options are management's initial assertions of leadership. They are by no means ironclad or well delineated. To get them in shape requires phase three.

Phase Three: Detailed Designs and Hard Choices

So how does Kiddieville, or a company similarly stranded in mediocrity, choose a discipline from among the options developed during phase two? It uses tiger teams. Tiger teams are small groups of high performers mandated by the executive team to turn realistic options into practical solutions. They comprise the company's best and brightest, the people who expect to run the company one day and have a deep interest in assuring that it will be around for them to run. Each tiger team is chartered to consider one of the viable options and thoughtfully answer the following questions:

■ What does the required operating model look like—i.e., what are the design specifications for the core processes, management systems, structure, and other elements of the model?

■ How will the model produce superior value?

■ What levels of threshold value will the market require in the other dimensions? How will these be attained?

■ How large will the potential and captured market be for this value proposition?

■ What is the business case—including costs, benefits, and risks—for pursuing this option?

■ What are the critical success factors that can make or break this solution?

■ How will the company make the transition from its current state to this new operating model over a two- to three-year period?

Parallel teams, working on a few competing options, give executives the luxury of comparing and contrasting the possibilities for the future. The teams give Kiddieville's executives several clear choices.

Executive leadership ultimately comes down to making the hard choice of a value discipline—what the company will stand for in its market and how it will operate to back up its promise. The decision to select a value discipline commits a company to a path that it will remain on for years, if not decades.

If after considering the options, Kiddieville executives choose to pursue shopping convenience as the company's distinctive value, then they have made significant implicit decisions about the organization's structure, management systems, business processes, and culture. An operating model for unmatched service convenience will closely follow the operational excellence model we earlier described. It will focus on the core processes of customer service and end-to-end product delivery. Employees will be regimented, disciplined, and directed. Systems will be used to drive zero-defect service convenience. For instance, Kiddieville will be among the first to invest in next-generation point-of-sale technology—if that technology can substantially reduce check-out delays.

If Kiddieville management chooses another option—to become a product leader in preschool toys and equipment, for instance—they will have made entirely different implicit decisions about the company's structure, management systems, processes, and culture. The focus would be on obtaining unique merchandise that represents the best in each category. The firm would build a deep understanding of how consumers value current merchandise; it would work closely with manufacturers on design; it might even create and control some of its own designs for products. The sales staff in a company that aims at product leadership needs less discipline and control and more time to sell the benefits of unique or unusual products. In-store displays and demonstrations also become very important for explaining the superior benefits of leading merchandise. Controls, structure, roles, and skills all vary according to the unmatched value Kiddieville would have committed to deliver. This third phase of activity doesn't challenge an executive team's alignment or creativity. Instead, it challenges them to show unprecedented courage. Does the executive team have confidence in its own analysis and the ability to commit to a particular course of action?

Courage in the face of doubt is essential because selecting a value discipline is not just a choice about what to do, it is a choice about what not to do—about what and who to leave behind on the journey toward market leadership. These are painful decisions, because management teams typically don't like to narrow their focus. But failing to focus, failing to choose one discipline and stick to it, is exactly what leads firms like Kiddieville to a state of mediocrity. The courageous will make decisions to get back on track. The cowardly, shrinking from the task at hand, will forever live with the memories of derailment, of painful journeys never completed.

1 1

CREATING THE CULT OF THE CUSTOMER

CHAPTER 11

CREATING THE
CULT OF THE
CUSTOMER

Some companies have it. Some companies don't. The "cult of the customer" is what we call it. You can't help noticing it when you find a company that has it, and yet you can't immediately put your finger on what it is. The cult of the customer is expressed in people's attitudes and behaviors. It's what separates the winners from the losers in the race for market leadership. Few market-leading companies—in fact, we know of none—have achieved or retained their position without a palpable culture that aligns precisely with their value commitment to customers.

HOW YOU RECOGNIZE IT

Inexact in concept, perhaps, the cult of the customer in practice takes on hard edges that are easily recognizable. Walk around a market-leading company and stop some rank-and-file employees in the hallway. Ask them what success means in their company and what makes them proud to be there. They'll probably talk about success in terms of value created for customers. Their pride derives from being able to touch their customers—directly or indirectly, but always tangibly. People throughout these organizations know that they can make a difference to buyers and users of their products and services.

Keep talking with the market leaders' troops, and you'll hear some deeply-held beliefs—call it the customer credo—that, in one way or another, captures two notions: First, customer value is the ultimate measure of one's work performance. Second, improving value (and the pace at which this is done) is the measure of one's success. Once absorbed into the fabric of a company, this credo assures that all employees engage their heads, their hands, and—in a sign of pure commitment— their hearts in going the extra distance for the customer.

The customer credo differs among companies pursuing different value disciplines. In operationally excellent companies, what stands out is employees' dedication to total dependability. Employees are proud that customers can count on them—that they're rock-solid in following up on what they promise. They want to be measured by how well they meet customers' expectations. The FedEx driver who was bent on delivering a package by 10 A.M., in spite of snowstorms and other daunting setbacks, typifies that attitude.

In product-leadership companies, the cult of the customer shows up in employees' missionary zeal to dazzle the customer with exciting new products, and to get customers to appreciate the value (performance, uniqueness, greatness) of those products. Or it shows up in employees' passion to create totally new customer benefits.

In customer-intimate companies, it shows up as a genuine interest in being the customer's trusted confidant and adviser. Employees go out of their way to help customers improve their business or squeeze more value out of the products and services they buy.

WHY IT IS IMPORTANT

In many subtle ways, market leaders weave their idiosyncratic, cult-of-the-customer philosophy into day-to-day management. Their customer credos create for employees a sense of common destiny, a deep understanding of what they're doing and why, and a sense of accomplishment. Leaders want each employee to take responsibility for doing what has to be done and accepting accountability for his or her actions. The delivery of customer value is a clear goal and can be used as an acid test for individual and collective success: Am I (are we) having an impact? Is my (our) work worth doing? Is there more important work for me (us) to be doing?

Without creating the collective purpose that the cult of the customer embodies, it is impossible for a company to become a value discipline leader.

A powerful cult of the customer is a potent antidote to the poison spread by whiners and passive complainers. Market leaders don't tolerate whimpers of "Ain't it awful?" and grumbles like "This company is all screwed up," or "If they'd only listened to me."

How You Get It

Getting to the state where people live by the credo requires a broad-based and ongoing effort to direct people's mindsets and behaviors. To lay the groundwork, market leaders don't wait until they've worked out all the details of their value propositions and operating models. As soon as they've settled on the discipline that will be at the heart of their business, they begin to raise the troops' receptivity to the new customer credo. In effect, they fertilize the organization for the value discipline that will soon be planted.

The building of the cult of the customer, like all broad-based organizational change efforts, follows well-established principles. For starters, people have to understand what the credo means in practical terms. Market leaders demonstrate this by communicating their value proposition and its implications for the company's operating model in a crisp, easily accessible manner. They lose no opportunity to repeat the basic precepts of what the company stands for and what's expected of its people. Wallet-size cards, ubiquitous posters, and desk-top paraphernalia constantly remind people of the company's deeply-held beliefs. Such trinkets may look corny, but to market leaders they exemplify resolve.

The next step is getting employees not just to understand the credo but to embrace it. Aspiring market leaders want the credo to pull on the heartstrings of their people. They want employees to wake up in the morning with a sense of the day's importance. They want people to know what they will do that day to create customer value—not just what they'll do to get through the day.

To achieve emotional buy-in, market leaders search for the right operational levers to pull. For instance, they go to great lengths to hire (and retain) people who enjoy working in their particular environment. They reward and recognize individuals' contributions in a manner that plays to their needs. They create an environment that lets people excel. And the impression they give through messages, symbols, and actions—

large and small—serves a similar purpose. A product leader's lavish offices and espresso bars project a different image than the frugal and efficient workplaces of an operationally excellent company. And customer-intimate employees are probably more concerned about having a workspace in their client's place of business (and being on the client's in-house voice-mail system) than they are about office space at their own headquarters.

The final principle in getting employees to live the customer credo is to accelerate action. That means removing obstacles that make it difficult for people to do what they rationally and emotionally know they need to do. It means giving people the tools and resources they need to work effectively and efficiently.

THE CUSTOMER CREDO
AND THE QUALITY MOVEMENT

In creating the cult of the customer, many market leaders astutely borrow from and build upon the quality movement. That movement's basic tenet is that work should be viewed as a process amenable to systematic streamlining and continuous improvement. Several principles support that position. One of them is "focus on the customer" (and customer expectations) in order to define the desired output of the process. Another is "prevent, don't fix," an admonition that drives strict adherence to norms and tested work procedures. Yet another is "measure everything," ensuring that facts and logic are used to guide process improvements and that workers have incentive and motivation. All of these principles, often tacked to bulletin boards in quality-minded companies, acknowledge the necessity of involving employees as the main contributors to and inspectors of work processes.

Market leaders know that the underlying logic of these quality principles is unassailable. Yet recognizing how quality principles have, alone, failed to guide some market laggards out of the doldrums, market leaders know that the quality movement is not the whole answer to achieving market leadership. So market leaders are selective in how they apply these generic management philosophies to the three value disciplines. In particular, the thinking of quality management falls short in meeting the comprehensive needs of the customer credo:

■ The customer credo focuses not on satisfying customers' expectations, but on delivering unmatched value—not just any value, and not just to please any customer. The credo focuses on pleasing the kinds of customers the company has decided to serve with the value those customers want.

■ Although many companies apply quality thinking to all operating processes, in reality it's been more successful in processes that are guided by explicit expectations, have measurable outcomes, and involve structured work.

Manufacturing and basic customer support are examples. Quality thinking has been harder to apply to less-structured work—such as marketing, product design, and information systems. Since mastery of any specific dimension of customer value encompasses multiple processes—some structured and some not—quality precepts have limited applicability.

■ Similarly, whereas quality management aims for continuous improvements across the board, the customer credo focuses on those areas where customer expectations can be raised and value propelled to new heights, either through ongoing improvements or periodic breakthroughs. This more precise focus requires that employees keep an open mind and an eye peeled for future, not just current, customer expectations.

These differences between quality and the cult of the customer suggest that market leaders need to adopt a slightly different set of principles. Tacked on their bulletin boards might be the following: "Tune in to value," "Live with the customer," and "Act as the customer's advocate."

TUNE IN TO VALUE

Tuning in to value starts with individual employees. The idea is to get each person to pay attention to how well he or she creates the kind of customer value that is consistent with the company's value proposition. Every employee needs some device—like an individual scorecard—that helps him or her make self-assessments in four central areas:

■ How does my work create value for the customer? Does it contribute to my company's value proposition? (If not, or if the answer is unclear or unsatisfactory, the employee should ask whether their job should exist.)

■ Am I delivering more value this year than I did last year? What accounts for that difference—or the absence thereof?

■ What could I do differently in the next year that would allow me to contribute greater value to customers? How could I help colleagues increase their own contributions to customer value in the next year?

■ What stands in the way of my doing better in the next year? What's needed to overcome these obstacles?

By routinely asking themselves this or a similar set of questions, employees have a basis for self-review. The same questions then form the foundation for discussions with their colleagues—peers, superiors, or subordinates—to ensure that they know what's important, how they're tracking against it, and how they can work together toward the common goal.

Market-leading companies supplement such self-assessment with systematic, objective monitoring of the company's overall performance. They adopt the total quality management principle of getting the facts required to scrutinize their actions. They zero in on customers' expectations and perceptions, on the performance of their vital operating processes, and on the practices of both competitors and leading companies in nonrelated industries.

With these objective measures of individual and collective performance in hand, a company can strengthen its cult of the customer. Management can, for instance, recognize that it's not just money that drives people. Public recognition of the right kind is a potent motivator, and it allows market leaders to make a statement about what is important in their organizations. In operationally excellent companies, the highest form of recognition may be bestowed on the best team player. In product-leadership organizations, it's individual feats that get celebrated. In customer-intimate companies, the highest honor is likely to go to those employees whose clients rave about their contributions.

The flip side of rewards—reprimands—are also parceled out with discrimination. Operationally excellent companies don't shy from penalizing team members when the team violates company-wide standards. Product leaders aren't reluctant to demote individuals whose performance is less than outstanding. To a great extent, the hero-culture of some product leaders self-regulates people's behavior by treating below-average performers with indifference. At customer-intimate companies, under-performers get the message when they're assigned to the least desirable client.

A telling characteristic of market-leading companies is that their senior managers consistently display their own devotion to the customer credo. To tune into value, employees need only watch the boss in action. The story is told of Bernie Marcus, head of Home Depot, who walked into the back office of one of his stores and noticed a Sears Craftsman wrench lying in a pile of items that customers had returned. Marcus called the store's customer-service people together, held up the wrench, and asked who had accepted it as a return, since Home Depot doesn't sell Sears wrenches. One employee admitted guilt, whereupon Marcus broke into a grin. This was a great example, he said, of someone taking responsibility for doing the unorthodox to please a customer. Never forget, he told the people, to go the extra distance for the customer.

Good management demands good role models. Most market leaders have their equivalents of Bernie Marcus—not just at the top, but throughout the organization.

LIVE WITH THE CUSTOMER

Thousands of managers sit in thousands of offices, content with the idea that they can know all about their customers without much effort. They think they can spend a day or so every year touring their customers' plants. Or that they can know all about consumers by chatting up a few of them at the checkout counter. Some think they can get a feel for customers by riding the company's delivery truck, reading through market research reports, digesting periodic customer surveys, or sitting in on the occasional focus group.

Not so. That's not how upstart AT&T Universal Card Services rose to challenge the giant card issuers. Nor how Southwest Airlines became the dominant carrier in the nation's busiest air-travel markets.

To their discredit, most managers are far too casual about inquiring into the minds and guts of their customers. Incidental and intermittent excursions are inadequate. They distort managers' views of customers with what social scientists call "anecdotal evidence." If most of the customers they see on any particular occasion seem to be happy, they're inclined to conclude that most customers are happy—which isn't necessarily the case at all.

Managers in companies that really succeed, in contrast, "live with the customer." What's different with these companies is that they make a continuous, painstaking, and unrelenting effort to get everyone to walk in the customer's shoes—to experience what the customer experiences, to feel the way the customer feels. Living with customers raises employees' sensitivity to their own products and services, not just at the time of purchase but after the newness has worn off.

Living with the customer is not a one-time or episodic event; it becomes a routine and substantial part of people's work schedules. They make it their business to know both their easy-to-please and their tough, demanding customers. In fact, they know that it's the latter who often challenge the organization the most, and in doing so suggest new ways the company can improve its value.

Living with the customer is a fact-laden experience. Ask the people at AT&T Universal Card, which requires every manager in the company to listen in on two hours of customer-service telephone calls each month. Or ask the managers at McDonald's. They go to great lengths—both at headquarters and in the field—to solicit customers' feedback. They track and dissect the opinions and experiences of hundreds of thousands of customers every year. And they act on that information, incorporating it in every decision in their endless quest to improve the customer's eating experience.

Living with the customer means going beyond the people in marketing, sales, or frontline customer service. Everybody in the organization—at workbenches and workstations—gets deluged with information and feedback on how their work affects what customers value. One leading vehicle manufacturer uses a systematic approach—it calls it value management—to codify different vehicle buyers' experiences. The codified information is used by the engineering staff to balance and rebalance design constraints—weight, cost, complexity, performance, aesthetics—against customers' desires and expectations.

It's still important, though, that employees remain focused. They don't want to live with just any customer, only with those to whom their company's value proposition is pitched.

ACT AS THE CUSTOMER'S ADVOCATE

Even employees in a highly competent workforce can't find easy solutions to every problem. Customers may make demands that are unprecedented. Employees may see opportunities for better products or services that nobody has seen before. What to do? If the answer can't be found in a company's rulebook, or if current practices appear to be inappropriate to deal with the situation, what then?

It's simple: You just ask people to use their heads. Give anybody—on the front line or in the lab—license to take on the role of customer advocate. Encourage and equip them to tackle issues affecting value creation reasonably and without a fuss. But be sure they understand that the customers for whom they are advocating should be the customers that fit the company's value discipline.

The general prescriptions that market leaders follow in getting employees to act as customer advocates are straightforward. They build employees' competence. They open up lines of communication between people in the organization whose cooperation is pivotal in enabling or expediting customers' wishes. They put information at people's fingertips, and they create fast feedback so employees find out quickly what works and what doesn't. They encourage employees to learn from their efforts, and, of course, they provide employees with moral support, confidence, and encouragement in their efforts to move the value frontier forward.

Customer advocacy is both a present and a future activity. Market leaders are clearly interested in filling the needs and expectations that customers have today. They are just as interested in responding to customers' dreams and wishes—that is, tomorrow's expectations about value. How companies respond to future expectations depends on the value discipline they're pursuing.

Operationally excellent companies encourage change in their operating procedures rather than one-time acts of heroism. It makes little sense for them to get employees to pursue out-of-the-ordinary customer needs. Rather, these companies are interested in creating communication vehicles that will keep feeding them customer information which they can then turn into systematic improvements.

Toyota, for example, encourages employees to submit ideas and suggestions. The company has been phenomenally successful in generating response (20 million ideas in 40 years—or several dozen per person annually) and in getting the vast majority of them implemented. Toyota's secret? It established a procedure in which employees prescreen their ideas (i.e., ask themselves whether they are focused on the credo of the company). Then it sends the screened ideas through a systematic procedure to mine for the ideas with merit.

In product-leadership companies, customer advocacy can involve championing new product concepts on the customers' behalf, lobbying to get a less-than-stellar new product improved or pulled from the line-up, or working with customers to find out what would constitute a leap forward in product performance. At Silicon Graphics, product developers and designers work together with leading-edge customers to define and design superior products. At a leading sound equipment company, marketing managers join engineers on flights around the world to visit people in all the offices of a large multinational customer. They help all the customer's people—not just its headquarters staff—in creating the specs for their new product.

Customer-intimate companies are by definition in the business of advocating the customer's interests. Employees in these companies are role models for customer advocates, since their daily work requires immersion in the customer's problems. For them, heroics on a customer's behalf are routine.

The cult of the customer does indeed defy a succinct definition. But you'll know it when you see it, and you'll miss it when it's gone. The cult of the customer is more than an abstract concept. It's part of the way market leaders do their daily business.

1 2

SUSTAINING
THE LEAD

CHAPTER 12

SUSTAINING THE LEAD

Pictured standing in a canoe on a 1986 cover of *Fortune* magazine was Kenneth Olsen, founder and then-president of Digital Equipment Corporation. He won the coveted cover spot for having piloted his corporate ship to record profits and growth. Yet just a few years later, Olsen was toppled by his board of directors into the cold waters of forced retirement. Was *Fortune*'s adoration the kiss of death?

Olsen was not the type of person to let publicity go to his head. His downfall, however, pumped new life into an enduring bit of management folk wisdom—that the moment an admired executive's face splashes across the covers of leading business magazines, the decline of the company he or she heads has begun.

Over and over, companies slump within a few years of a rave review. It's happened to IBM, Westinghouse, American Express, and Kodak. The kiss of the media turns into a curse.

Maybe there's no established cause-and-effect relationship between celebrity and decline, but the hangovers suffered by companies that drink the heady wine of fame are too common to ignore. That yesterday's stars so often turn into today's has-beens can't stem just from the evil eye of fate. We believe that praised and praiseworthy companies often fall into decline because they fail to maintain their well-conceived strategies. Having attained market leadership, many firms celebrate their victory, admire their own operating model, and exploit their advantages for shareholder gain. They simply rest on their laurels. And as they recline in the warmth of adoration, they fail to see they are violating a central rule of market leadership: Dominate your market by improving value year after year.

Sustaining market leadership is a full-time job. Unless the energies of a company are fully mobilized to continuously create major improvements in value, it is impossible to retain a lead—especially when every competitor is hungrily working to knock off the leader and claim the top of the hill for itself.

But staying focused and committed to customer value is hard. Success offers such tempting distractions and so many excuses for delaying improvement.

TO STAY AHEAD, YOU HAVE TO GET BETTER AND BETTER

Companies that expect to escape the fate that toppled the likes of Digital must accept business as a copycat world. Little in business stays proprietary for long—not products, processes, technologies, nor strategies. People emulate—if they don't actually duplicate—what wins. Almost all market leaders will find admirers and competitors copying their successes. Continental Airlines is trying to replicate Southwest Air's operating model. Microsoft is roughly matching Apple Computer's easy-to-use operating systems. And Consolidated Freightways' Menlo Logistics unit uses much the same approach to customer intimacy as Roadway Logistics Systems.

Copycats are on the prowl at all times in all industries. So how do market leaders of whatever value discipline stay ahead? They maintain the focus of their value discipline and intensively compete with their own success. They work continuously and simultaneously to improve their operating model and make it obsolete. They are operational excellence firms striving to reach entirely new benchmarks of price and hassle-free service. They are customer-intimate firms trying to make their own total solutions obsolete. They are product leaders trying to destroy demand for their current products with dazzling new ones. Better they should do it than the competition.

For every market leader, advances in value to customers are gained by tightening performance standards, reengineering work processes, and upgrading competencies.

For operationally excellent companies, the toughest challenge is to shift to the next generation of "no frills" standardized assets to achieve the next level of efficiency. For product-leadership companies, the toughest challenge is to see the next technology, the next concept, that is beyond the bounds of their expertise. For customer-intimate companies, the toughest challenge is to let go of current solutions and to move themselves and their clients to the next paradigm.

Let's look at each of these challenges in turn.

SUSTAINING OPERATIONAL EXCELLENCE

Some companies stress the application of efficiency-enhancing assets to such an extent that they overinvest in the current paradigm. Prosperous firms can well afford the capital outlays, so they try to buy their way to efficiency by, say, acquiring newfangled machines. Line managers can often sell finance chiefs on projects that might otherwise get a thumbs down as offering too little benefit for the cost.

American Express fell into this trap in the 1980s; its $1 billion Genesis program was an ambitious effort to boost operating efficiencies through information technology. Reportedly, AmEx collected large amounts of data on cardholders' buying patterns. It wanted to have the most advanced information technology. But the effort distracted AmEx from more pressing concerns, in particular the increasing price sensitivity of its merchant and consumer clients. This heavy investment in "efficiency" ultimately only added to AmEx's cost structure.

■ Threat: Assets That Turn into Liabilities

Market leaders can easily spend too much time figuring out how to better utilize assets that may no longer be the right ones. At American Airlines, the high-tech SABRE division made a science out of streamlining standard operating procedures, systemizing reservations, and filling planes. Yet American's core assets—namely planes, gates, and hub infrastructure—now block further improvements.

American's major hubs at Dallas/Ft. Worth and Chicago's O'Hare are high-cost facilities compared to Southwest's facilities at smaller airports in the same cities (Love Field and Midway). American's airplanes span a wide variety of makes and models. That's the kind of variety that adds complexity and kills efficiency. And the SABRE system is itself a high-cost asset, absorbing large annual investments in maintenance. American Airlines had the right kind of assets for the established operating model in the domestic airline business, but they were not the right assets for the hyper-efficient operating model with which Southwest has challenged the industry. Restructuring of such big, inflexible assets may be inevitable if American wants to rival the efficiencies enjoyed by Southwest.

Wal-Mart may yet stumble as it encounters similar changes. It has done a yeoman's job of streamlining operations. It anticipated and mastered the operational skills for running warehouse stores. Together with PriceCostco, it popularized and pushed to the limit the wholesale club. But a move by consumers toward home shopping, for instance, could turn a lot of Wal-Mart's assets—stores, distribution centers, long-haul trucks, inventory systems—into liabilities. Wal-Mart's standard operating procedures could become irrelevant. Should a mass migration to home shopping materialize, consumers would redirect their orders to home shopping networks. Shipments could go out UPS or FedEx. Customer billing could be handled by credit card. Wal-Mart managers, gazing wistfully at empty parking lots, would pine for the days of booming customer traffic.

MCI Telecommunications could face similar risks. Its main asset, its telephone network, enables the company to promise low prices on long-distance calls. But what happens when cable TV companies offer telephone service through their coaxial or fiber-optic lines? What happens as telephone, computer, and TV technologies merge? MCI may need to make big moves to restructure its asset base to cushion risks and grab new opportunities. If MCI doesn't move, or move soon enough, expect its star to dim by the prescience of companies like Microsoft, Time-Warner, and Tele-Communications, Inc., which are currently investing in this still-unexplored future.

SUSTAINING PRODUCT LEADERSHIP

Product leadership companies, in particular, become fascinated with great products. As a result, with each bright new idea or blunt complaint, the developers rush back to their labs in a perfectionist panic. An assessment of whether the perceived shortcomings are relevant or not is neglected in the rush.

Getting too close to their customers can distort people's focus. For instance, high-tech companies frequently set up so-called user groups to critique products. When the engineers get their most enthusiastic and sophisticated users in a room to share experiences, a deluge of complaints and suggestions results. The engineers can hardly ignore the priority demands, but can they afford to accept them? At such times, product leaders must keep their goals in focus.

Sustained product leadership comes only from a deep commitment to breakthrough innovations. Those ideas aren't usually gathered at user group meetings, because users want to hone and polish the product they already have. Customer feedback is important; it helps to improve the value of existing products and to extend their life. But if investments in enhancements impede progress toward breakthroughs, then a serious mistake is made. Product leaders must make their own products obsolete; they must compete with their own success if they are to sustain their lead.

Product leadership companies are prone to stumbling and fumbling when fundamental technologies and market conditions change radically. Experts, especially technical experts, are particularly prone to this human frailty. In 1895, a mere eight years before the Wright brothers' historic first flight, Lord Kelvin, president of Britain's Royal Society, the most prestigious scientific group in the world, declared that heavier-than-air flying machines were impossible. In 1920, Robert Millikan, a co-founder of quantum mechanics, declared that there was no likelihood that man would never tap the power of the atom. In 1899, Charles Duell, head of the U.S. Patent Office, observed that everything that could be invented had been invented! Jack Warner, co-founder with his brothers of Warner Brothers studio, which produced the first talking picture, was skeptical of the new medium. "Who the hell," he asked rhetorically, "wants to hear actors talk?"

■ Threat: Sense That Turns into Nonsense

Not only assets soar or plunge in value as markets change. Ideas come and go as well. Yesterday's truisms become tomorrow's falsehoods, and a company can easily develop blind spots that impair its people's skill at sensing the potential of new technologies and concepts.

Remember General Motors in the 1970s and early 1980s. GM didn't understand or want to act on the changing tastes of young car buyers. When they wanted lighter, faster, nimbler cars with a European feel,

GM's sense of that market segment resulted in nonsense: GM continued to believe that Chevy buyers would upgrade to bigger Buicks, and ultimately to colossal Cadillacs. Old truths had changed, but GM hadn't changed its thinking.

Who knows what will happen to cars next? The green movement has prompted Toyota to plan for environmentally-sound, fully-recyclable cars. Upcoming regulations governing fuel economy, material recycling, and employer-mandated car-pooling to cut urban smog could shake up the car industry even more. Will we see an explosion of battery-driven vehicles? Or computer-guided transportation? Product leadership firms in the automobile industry don't know the answers to all these questions, but they must be exploring them so that they can stay ahead of the pack.

Market changes put heavy pressure on product leaders' sensing skills. For the U.S. drug industry, for instance, markets have shifted toward generics. This shift puts to the test many of the procedures at companies such as SmithKline Beecham and Merck, which previously developed and merchandised products protected by their specific name value. A further challenge for drug makers is genetics research, which focuses on preventing disease, not curing it. For example, by the end of the decade, it's expected that researchers will discover genetic markers for the four kinds of cancers that account for most cancer deaths. These discoveries could make certain traditional drug therapy and treatment obsolete—turning what makes sense today into nonsense tomorrow.

If companies aren't going to succumb to myopia as their markets change, they must invest in building the sensing skills that will help them know when a product is ripe for the market or when it is due for retirement. At the same time, however, product leaders must continue to speed innovation if they expect to sustain their lead.

High-tech product leaders such as Microsoft invest heavily to accelerate invention—partly through close ties with faster-moving start-up firms. Media companies such as MTV are striving to stay a step ahead of a society that is evolving at a dizzying pace. Some product leaders are reengineering for speed. Pharmaceutical firms are reducing the time it takes to market their products by working with the Food and Drug Administration to redesign its drug-approval process.

Consumer product leaders are bringing concepts to market faster by simulating them with software or testing them in market labs called "greenhouses."

SUSTAINING CUSTOMER INTIMACY

Customer-intimate companies are particularly susceptible to the illusion that they can do absolutely anything to give customers the total solution promised. It leads them to take on tasks they should decline or should pass on to other suppliers. For instance, they may persist in providing services they once performed uniquely well but which over time have been copied by so many competitors that they've become commonplace. When yesterday's premium services become today's basic standard, the customer-intimate company has to find ways to offload them.

■ Threat: Knowledge That Turns into Ignorance

All too often, the blast that knocks a high-flying company out of the sky comes not from a competitor with stronger sets of the same skills, but from one armed with skills that the high-flyer doesn't have. As the old saying goes: It's not what you know, but what you don't know, that hurts you.

One simple example: In the 1970s, IBM found that its dedicated customer base of information-systems professionals had shifted roles. The technology mavens were no longer the main specifiers and buyers of IBM products. IBM found itself ignorant of the needs of its new clientele, the financial executives and line managers who now signed the purchase orders. Like seasoned businesspeople set down in a foreign land, IBM's sales force knew the business but not the new language they were expected to speak.

Cott Corporation, to pick just one customer-intimate market leader, has to worry about the future in lots of dimensions. Giving customers the soda formulation and packaging they want may not be enough to retain clients. Wal-Mart may want Cott to tie in with its sophisticated cross-docking procedures. Safeway may want it to tie in with leading-edge store-door deliveries.

Does that mean that Cott's network should include a world-class logistics provider? Should Cott forge alliances with firms that conceive and build such capabilities? How knowledgeable does Cott itself have to be to excel in brokering these co-providers' services? Could its expertise in one area be threatened by its ignorance in others? Those are the kinds of questions customer-intimate companies have to think about continually in order to sustain their leads.

Cott's situation shows how customer-intimate companies' delivery systems can make or break them. Whether Cott delivers the total solution itself or draws on a network of co-providers, it nonetheless receives all the blame or credit.

The other distinctive asset possessed by customer-intimate firms is their consulting expertise. Customers depend on it, but since the value of such expertise diminishes as customers themselves become more expert, customer-intimate companies must learn to stay two steps ahead of customers. They must assimilate experience from multiple clients, acquire fresh insights by hiring new people, and pick the brains of expert outsiders. To beat back rivals, they have to adapt their expertise to new clients and to the changes in existing clients' basic problems. Staying smart is their greatest challenge.

MAINTAINING THRESHOLDS OF PERFORMANCE

Of course, even if market leaders stay ahead by constantly refining those capabilities that keep them on top, they can still fall behind if they don't periodically improve secondary disciplines. Minding these is most critical when competitors are resetting customers' expectations for performance. For instance, as Ford and Toyota elevated expectations about automobile quality, they blocked Hyundai from making significant market inroads even though Hyundai products were bargain-priced. Dell Computer and Gateway 2000 reinvented competition in the personal computer industry by focusing on price and convenience, which had been secondary disciplines for the big, entrenched computer makers, who then had to improve their own performance in these dimensions.

One product leadership company hit hard by rising customer expectations in price and convenience was Apple Computer, which responded with improvements that enabled it to drop prices on some of its products by as much as 34 percent and meet the resulting surge in customer demand. Apple may still have some way to go to catch up with rapidly rising operational standards in the PC business, but it has at least shown that it's capable of a response. Digital Equipment Corporation, on the other hand, was caught flat-footed.

Digital was an impressive market leader until the early 1990s, riding high on its reputation as product leader. But the company lost its product edge, and then it plunged into the red by underestimating its deficiencies in operational performance. At the end of 1990, the company had 50,000 more employees than it could afford, and even then it could barely match IBM's unimpressive productivity level as measured by sales per employee.

Executives should be wary, however, of overzealous efforts to polish their company's secondary disciplines. Too much attention paid here can deflect attention from the more important tasks required to strengthen the company's value proposition. The goal is to sustain threshold levels of the other values, but not overinvest. In the early 1980s, McDonald's, for example, overreacted to competitive innovations by loading up its menu with too many new items—pizza, tacos, chicken fajitas. These items created confusion and complexity in the operation without undercutting the threats posed by specialized pizza parlors or Mexican food purveyors. The company had to retrofit kitchens to prepare the new foods. It had to train suppliers to handle different orders. Worse yet, its 18 million daily U.S. customers got confused about what McDonald's stood for. Whatever happened, customers wondered, to the good old Big Mac? McDonald's quickly came to its senses and recommitted itself to keeping its eye on the ball it could hit best—operational excellence.

General Motors also went astray in the early 1980s when, bent on bolstering its operational excellence, it poured money into plant modernization and innovation. Some new investment was essential, but then-chairman Roger Smith went overboard. He sought, through billions of dollars of investment, to build the factory of the future, to catapult the company way beyond the state of the art in manufacturing technology. He aimed to substitute robotics for people in order to drive down manufacturing costs. Smith's overweening faith in technology and in his company's ability to harness it mired GM in challenges far beyond its capabilities. GM, in effect, took itself to the cleaners.

NARROW FOCUS THAT TURNS TO MYOPIA

Try as they might to retain operational focus, market leaders sometimes lose it. The leaders' vision becomes blurry, not from gazing at a far horizon but from staring at the ground before them too closely. What Polaroid valued most and did best was research, but instead of allocating energy and resources to exploiting the results of that research, it just continued to do more of it. It became myopic.

More commonly, companies grow short-sighted because they're tempted by short-term rewards. They think that squeezing more money out of an existing business will, for instance, boost their stock prices: "Wall Street will love us." Other companies believe their strong brand name will let them raise prices with impunity or that if costs come down, they can bank them instead of passing them along to the client.

Cereal makers like Kellogg—whose products now cost more than Casio's calculators—will steadily lose market share. Wal-Mart, on the other hand, has resisted the temptation to inch its prices upward and has been rewarded by steady, impressive growth.

Expansion that's too rapid is another form of corporate myopia. Toys "R" Us concentrated on global expansion at the expense of running its existing stores, which gave Wal-Mart the opening it needed to grab an increasing share of what had been the Toys "R" Us market.

Both Wal-Mart and Toys "R" Us are classic operationally excellent retailers that emphasize low prices and hassle-free basic service. Wal-Mart leveraged its $80 billion scale to take over as the price leader, though Toys "R" Us's prices are still comfortably lower than most other specialty retailers'. Toys "R" Us offers the broadest selection in the industry, but Wal-Mart has started to close the gap during the crucial Christmas selling season. It's no contest on location convenience: Wal-Mart has four times as many locations coast to coast, and many consumers are in the habit of regularly shopping in these stores. As for in-store experience, Wal-Mart has trained its staff to be friendly and helpful without giving up much efficiency. Toys "R" Us staff is poorly trained and often indifferent to customers' needs. What's been the result? Wal-Mart has almost doubled its market share over four years to 16 percent. Toys "R" Us has grown only modestly, to about 22 percent. Toys "R" Us needs to improve its value proposition if it's going to stay at the top.

In the late 1980s, IBM gave out biased advice when it told clients that it could meet all their computing needs, even as the marketplace began to favor supplier-neutral "open systems" and the use of multiple vendors. Today, IBM is trying to heal the wounds it inflicted on its own reputation. IBM account reps now explore both IBM and non-IBM possibilities when advising customers how best to meet their specific needs.

Market leaders place their leadership in jeopardy when they exploit value at the expense of innovation. Pepsi, for some reason, believes that dating its soda cans will generate more customer demand, and Coke believes that reintroducing in plastic the corset-shaped bottle that helped make it famous will rejuvenate the brand. Meanwhile, Snapple and other aspiring leaders are making sharp inroads in the soft drink market. Their "secret": they're doing real product innovation instead of just tinkering with their package design. Compare their effort with that of pharmaceutical maker Glaxo, which has invested in 200 outside research programs that could potentially feed new drugs into its pipeline.

The greatest temptations facing market leaders? Getting greedy, milking their success, and not moving forward. Operationally excellent companies can get tempted into overpricing, product leaders into underinnovating, and customer-intimate firms into underservicing.

Balance is essential in all three value disciplines for sustained market success. It leads to win-win outcomes that generate payoff for everyone: customers, employees, and shareholders. Imbalance, unless corrected, pitches companies into a death spiral: Increasingly less profit forces them into further value exploitation, which accelerates the downward plunge.

Each one of the three value disciplines has its own strategy from which a company diverges at its peril. That's not to say that tactics can't change; a successful company is infinitely flexible in acting to sustain its lead. But myopia, temptation, the loss of balance, and other such perils can mortally wound the strategy that sustains leadership.

EPILOGUE

It's conceivable (indeed plausible given current trends and patterns) that business historians in the year 2001 will look back at the last decade of this century as an era of profound change—perhaps even a historic turning point.

During that fateful decade, historians may conclude, a new breed of market leaders emerged. These leaders weren't simply the younger, more vigorous offspring of the old breed; they didn't just work harder to meet tougher customer demands. Instead, they broke the mold and built a new one—by basing their success on one of the disciplines chronicled in this book. They chose their customers deliberately, walking away from those that didn't fit their visionary new mold. They chose to narrow—not broaden—their operational focus—"all things to all customers" suddenly seemed as outdated as a manual typewriter. And by committing themselves to deliver more value year after year after year, they were able to sustain their lead and dominate their markets.

To turn their determination into deeds, they upgraded, renovated, and built new operating models capable of delivering unprecedented levels of customer value, which in turn raised customer expectations. Skyrocketing expectations raised everyone's threshold—the new leaders had redefined the concept of competition.

During the 1990s, our business historians saw operating models evolve at a dizzying pace—first at the hands of the pioneers discussed in this book. They were soon followed by a wave of struggling companies bent on emulating the new leaders' success. Then came the newcomers, the startups, the entrepreneurs eager to capitalize on the innovative new operating models.

As established companies narrowed their focus and learned to master the discipline of market leaders, they became aware of the constraints of their existing, traditional business structures. This awareness led the larger, integrated corporations to restructure themselves into separate units that each excelled in a single discipline. Shared resources were kept to a minimum.

Companies also turned to other companies for help in designing and running those parts of their operating model that were necessary—but not critical—to value creation. This, in turn, led to a surge in demand for new corporate connections in the form of outsourcing, joint ventures, and strategic alliances.

The business historians will look back at the 1990s as an era in which the workforce was reenergized. The new insights gained about value creation spurred innovation in products, services, and customer relationships. They provided much-needed relief from the energy-zapping and demoralizing downsizing that had characterized corporate behavior in the early years of the decade. The mid- and late-1990s were a period of optimism and renewed purpose for companies and their employees.

The business historians will marvel that a simple idea—better value for customers year after year—had such profound ramifications. And customers? They just loved it (and kept expecting more and more).

INDEX

INDEX